# Lectures on Location Theory

Springer
*Berlin*
*Heidelberg*
*New York*
*Barcelona*
*Hong Kong*
*London*
*Milan*
*Paris*
*Singapore*
*Tokyo*

Martin J. Beckmann

# Lectures on Location Theory

With 36 Figures
and 4 Tables

Springer

Prof. Dr. Martin J. Beckmann
Brown University
Economics Department
64 Waterman Street
02912 Providence, RI
USA

ISBN 3-540-65736-3 Springer-Verlag Berlin Heidelberg New York

Library of Congress Cataloging-in-Publication Data
Beckmann, Martin J.
Lectures on location theory / Martin J. Beckmann.
p. cm.
ISBN 3-540-65736-3 (hc. : alk. paper)
1. Industrial location. 2. Space in economics. 3. Time – Economic aspects. I. Title
HD58.B319 1999
338.6′042 – dc21 99-25888 CIP


Printed in Germany



Hardcover-Design: Erich Kirchner, Heidelberg

SPIN 10721828 42/2202-5 4 3 2 1 0 – Printed on acid-free paper

**Robert Herman (1914–1998) *in memoriam***

# Preface

This textbook is short, because it does not aim to be comprehensive, i.e., presenting and explaining every line of approach that was ever tried on spatial phenomena by economists, geographers or social scientists. Rather, it sticks to one line of argument as closely as possible, following as far as possible established (non-spatial) microeconomic theory.

This permits one to try for consistency and systematics throughout. In this it resembles my earlier "Location Theory", (Random House, New York, 1968) of which, however, it is not a second edition.

My intent has been to bring order (in truly Germanic fashion) to a somewhat uneven and disconnected field which is known to abound in *ad hoc* reasoning and as lacking the rigor and charm of general economic theory. To the small band of location theorists it is, however, an object of great beauty which we are not adverse to share with others.

My friend and repeated co-author Tönu Puu has just published his own "Mathematical Location and Land Use Theory" addressed to the mathematically minded. The present book is cast at the more elementary level and written for general economists. Mathematical arguments are given in notes at the end of each chapter.

Now integrating location theory into neoclassical economics is not as straightforward as it may seem. For space brings with it two phenomena that raise difficulties and are therefore often ignored in neoclassical economic theory: economies of scale as the prime cause of "localization" and externalities or "neighborhood effects", the cause of so-called agglomeration.

Fortunately, the diminishing returns from transportation can eventually offset the increasing returns in production and thus make decentralization and spatial equilibrium possible. Agglomeration remains troublesome, however, as a cause of spatial concentration with negative welfare effects. We invite the reader to examine with us these challenges to conventional economic analysis.

I have learned from discussions with my friends Åke Andersson, Tönu Puu, Wei Bin Zhang, Yang Kai Feng and benefited from the stimulating and supportive environment offered by the Institute for Futures Studies in Stockholm.

My special thanks go to Mrs. Gunov Albihn, principal secretary of the Institute, without whom this book could never have been written. I also acknowledge with thanks the invaluable help given me by Mrs. Marion Wathey in the final revision of this manuscript.

The stimulus to write this book came from giving a lecture course on location theory in the Geography Department of Peking University at the invitation of Professor

Yang Kai Feng in the fall of 1995, an enjoyable experience because of the lively interests of my students. I wish to thank the DAAD and especially Dr. Gisela Schultz for supporting this "Kurzzeit Dozentur".

Providence, March 1999 Martin J. Beckmann

# Contents

## Chapter 15

# CHAPTER 1

# Interaction and Distance

Space, the two-dimensional surface of the earth, enters into economic relationships in two ways: first through the (uneven) distribution of natural resources and secondly through distance separating economic activities.

Natural resources, both of the exhaustible and renewable type, can attract mining, drilling and other extraction activities, and under certain conditions become the location of further processing activities (chapter 5). In view of that, "natural harbors" should also be included among the natural resources.

The "renewable resources" are plants and animals grown on suitable land. Land capable of agricultural use is an important attractor of economic activities, to be treated under the general heading of land use (chapter 11).

In order to study the pure effects of distance, one assumes away all localized natural resources or places them all in one point, at the location of an emerging central city, an ingenious idea of Johann Heinrich von Thünen, the father of location theory (1826).

Two types of models thus characterize modern location theory, those focusing on the given location of natural resource deposits, and those ignoring them and assuming instead a homogenous plane, offering no special advantages to any activity initially. Then distance is the only spatial factor to be considered. The role of distance can be succinctly summarized in the statement, comparable in its universality to the "law of demand":

*Interaction decreases with distance*

We postpone a discussion of how distance should be measured and look at various types of interaction to which this law may apply.

## Interaction Observed

A systematic study of the effect of distance were first made in connection with *migration* which is perhaps the most drastic event in a household's economic existence and where distance therefore may loom largest. (Ravenstein 1885).

A short term analogue to migration is *commuting*, the daily movement between residence and workplace, that fills so many hours in so many lives.

Migration and commuting both involve transportation of persons, and transportation activities incur money and time cost and possibly discomfort and inconvenience. All of these rise with distance, and distance acts therefore as a deterrent.

In the case of migration another important cause surfaces: information. The greater the distance, the less is known about the potential destination. Migration thus often follows in the footsteps of family members, friends or acquaintances who have moved before.

Transportation is the interaction also in the making of *trips* for various purposes: shopping, school, church, family visits, recreational: weekending, tourism and extended holidays of the resort, camping, sports activity or "adventure" types.

An activity for which the distance effect is well-documented is *communication.* First studied were telephone calls (Carrothers 1956) and mail (Hammer and Iklé 1957), for both of which charges tend to increase with distance. But the law of decreasing interaction applies to all forms of communication, beginning with oral exchange.

The effect of distance on transportation and communication is a reflection of the underlying decrease of *human interaction* of such diverse types as romance and marriage on the one hand and scientific collaboration on the other (Beckmann 1994).

It has been shown that even *citations* in scientific publications decrease with distance, presumably because of the information effect (Beckmann and Persson 1996). Scientific collaboration, even when carried out by researchers in different locations, invariably requires face to face contact at some point, and this brings in the impediment of distance. Telecommunication will then facilitate the progress of work regardless of distance.

While "absence makes the heart grow fonder", a romance will hardly survive prolonged periods without face to face contact between lovers. The type of interaction most important for economic analysis is economic *exchange* in various forms: *trade* in general, newspaper and magazine *circulation*, bank *loans*, consulting and *service contracts* etc. These are usually studied under the heading of *market areas* (chapter 2).

The principal factors responsible for the decline of interaction with distance have been named already: the expenditure of time, money and effort and secondly the loss of information over distance. Interaction may also be deflected by "intervening opportunities" (Stouffer 1940).

A more subtle effect results from diffusion: contacts between persons or organizations who were in the same location at some former time and who have moved apart. In this class may also be found *college admissions*, when students seek out a parent's former school for their own study.

## Distance

Distance "as the crow flies" is straight line distance based on the Euclidean geometry of the two-dimensional plane. Other distance measures become economically relevant when interaction (mainly transportation) is channeled through the links of a given (transportation) network. A simple and fairly common type is that of a "square grid" with parallel roads of two sets that intersect at right angles, say a North-South and East-West road grid. Distance is then the sum of the East-West and North-South coordinate differences (in absolute terms). This is sometimes called "Manhattan metric".

In an arbitrary network distance between any two points is determined on "shortest paths" which may be discovered from data on the link length by the use of linear programming or dynamic programming algorithms (mathematical note 13.1). The human eye can often see a shortest path by inspection of maps.

Distance is defined by shortest paths. Unit transportation costs may vary among locations, say because of the terrain or of impediments to traffic by human settlements or traffic volume. When these variations are taken into account, shortest paths will not be straight lines any more. (In a continuous plane they are curves that are solutions to an "iso-perimetric problem" of the calculus of variations). The loci of equal distance from or to a given point are called "iso-vectures". They reach farther along corridors defined by special transportation facilities (super highways). When transportation costs are equal everywhere the iso-vectures form a set of concentric circles. If locally defined transportation costs are the same in all directions, then iso-vectures and shortest paths will intersect at right angles.

For many types of interaction the expenditure of time is more relevant than physical distance even when the latter involves a money cost. To find the most time saving shortest paths, delays due to congestion may have to be taken into account. To do this for all (potential or actual) road users calls for sophisticated programs of "traffic equilibrium analysis on capacitated networks" an important problem in transportation systems analysis (mathematical note 13.2).

In prediction, planning and economic analysis, one needs to know quantitatively how much interaction is affected by distance. This is studied in the following mathematical notes.

## Mathematical Note 1.1
## Alternative Interaction Laws

Let $A = A(r)$ be the level of the interaction activity and r the distance involved. Four possible relationships emerge from four alternative assumptions of the simplest type.

$$dA = -k\,dr \tag{1}$$

$$A = A_0 - kr \tag{1a}$$

a linear law.

$$\frac{dA}{A} = -k\,dr \tag{2}$$

$$\log A - \log A_0 = -kr$$
$$A = A_0\,e^{-kr} \tag{2a}$$

an exponential law.

$$dA = -k\frac{dr}{r} \tag{3}$$

$$A - A_0 = k\,\log r_0 - k\log r$$
$$A = \hat{A}_0 - k\,\log r \tag{3a}$$

a logarithmic law.
Notice that this applies only beyond a positive minimum distance $r_0$.

$$\frac{dA}{A} = -k\frac{dr}{r} \tag{4}$$

$$\log\frac{A}{A_0} = -k\log\frac{r}{r_0}$$

$$A = A_0\left(\frac{r}{r_0}\right)^{-k} \tag{4a}$$

a power law applicable beyond a minimum distance. In particular for

$k = 1$
$$\frac{A_1}{A_2} = \frac{r_2}{r_1} \tag{4b}$$

a reciprocal law,

for $k = 2$
$$\frac{A_1}{A_2} = \frac{r_2^2}{r_1^2} \tag{4c}$$

"Reilly's law".

In human interaction, at very short distances, interaction is almost constant; it will then fall off at accelerated rates and enter a convex regime with decreasing rates of decline. This profile can be captured by postulating

$$\frac{dA}{A} = -\frac{2r}{1+r^2}dr$$

or

$$\ln A - \ln A_0 = -\ln\left(1+r^2\right)$$

$$A = \frac{A_0}{1+r^2}$$

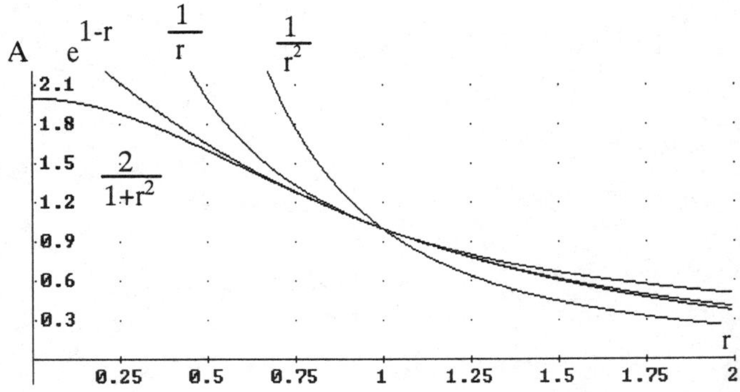


**Fig. 1.1.** Three distance laws standardized to $A(1) = 1$ (courtesy of T. Puu)

$$\text{which is} \begin{Bmatrix} \text{concave} \\ \text{convex} \end{Bmatrix} \text{for} \quad r \cdot \begin{Bmatrix} < \\ > \end{Bmatrix} \frac{1}{2}$$

which for large distances approaches Reilly's law $\frac{1}{r^2}$.

## Mathematical Note 1.2
## Gravity Laws

For predictions of behavior that is influenced by distance it becomes necessary to quantify its effect. It is tempting to draw on analogies to the decline over distance of forces in physics which usually obey an inverse square law (resulting from the three-dimensional space in which physical forces act). A "gravity law" for interaction means basically that interaction can be understood as the multiplicative effects of three separable factors: the generating capacity of the origin, the attraction of the destination, and the "resistance" incurred through distance

$$x_{ij} = Q_i A_j h(r_{ij})$$

$Q_i$ source or originating capacity
$A_j$ attraction of destination
$r_{ij}$ distance
$h(r_{ij})$ a resistance function

It is interesting how fashions have changed in the specification of the resistance function. In the early days of regional science "power laws" were favored by analogy with the laws of physical nature (Zipf 1941, Stewart 1950, Wartz 1959), while the "entropy school" spearheaded by Alan Wilson (1969) has brought in the exponential law to become the undisputed favorite on both theoretical and empirical grounds. Most interactions appear to obey at least approximately the same general type of "gravity law" with differences only in the parameters of the generation, attraction, and resistance functions.

$$x_{ij} = \gamma \, Q_i^{\alpha} \, A_j^{\beta} \, e^{-\lambda r_{ij}}$$

where source capacity and destination attractions are considered power functions of a "size" measure $Q_i$ $A_j$ and the effect of distance is stronger or weaker depending on the size of the $\lambda$ coefficient.

In a detailed analysis of particular types of interaction, alternative specifications of the distance effect should still be considered (see mathematical note 1.1).

## Mathematical Note 1.3
## Random Utility

To show how market shares are influenced by relative distance we introduce a "random utility model" (following Leonardi and Papageorgiu 1980). Let $u_1$ , $u_2$ denote the utilities of two competing heterogeneous commodities 1, 2. Net benefits to a consumer calculated as utility minus cost are then

$$u_i - p_i - kr_i$$

where $p_i$ is price k transportation cost per distance and $r_i$ distance between supplier i and customer. The commodity i = 1 is preferred wherever

$$u_1 - p_1 - kr_1 > u_2 - p_2 - kr_2$$

Now let utilities $u_i$ differ among customers at a given location according to a frequency or probability distribution which is the same everywhere. The probability of 1 being preferred is then the market share of commodity 1. Thus

$$m_1 = pr\left(u_1 - p_1 - kr_1 > u_2 - p_2 - kr_2\right)$$

or

$$\begin{aligned} m_1 &= pr\left[u_1 - u_2 > p_1 - p_2 + k\left(r_1 - r_2\right)\right] \\ &= pr\left[u_2 - u_1 < p_2 - p_1 + k\left(r_2 - r_1\right)\right] \end{aligned}$$

Let $\Phi$ be the probability distribution of utility differences. When $u_i$ is normally distributed, this difference is also normal. Now

$$\begin{aligned} m_1 &= \Phi\left[p_2 - p_1 + k\left(r_2 - r_1\right)\right] \\ &= H\left(r_2 - r_1\right) \end{aligned}$$

where H is an increasing function. This shows that market shares decline with distance, or rather with the differences of distances to suppliers.

When utilities are normally distributed and the normal distribution is approximated by a logistic, one has

$$\begin{aligned} m_1 &\doteq \frac{1}{1 + e^{-\frac{a}{\varsigma}\cdot\left[p_1 - p_2 + k\left(r_2 - r_1\right) - \mu_{12}\right]}} \\ &\doteq ce^{-\frac{ak}{\varsigma}\left(r_1 - r_2\right)} \end{aligned}$$

for large distance differences $r_1 - r_2$, an exponential distance law.

## Mathematical Note 1.4
## Gravity Derived From Utility

Travelers decisions just as consumer decisions in general should be derived from the economic principle of preference or utility maximization (Niedercorn and Bechtold 1969).

We consider a utility function whose arguments are frequency $x_k$ of visits to destination k, time available for other activities, and attraction $a_k$ of destination k for purposes of these visits.

$$u\left(x_k, a_k, t\right) \tag{1}$$

Let $r_k$ denote distance to k measured in time units, and let T be total time available so that

$$t = T - \sum_k r_k x_k \tag{2}$$

The utility function is assumed to be

$$u = \sum_k v(x_k, a_k) + w\left(T - \sum r_k x_k\right) \tag{3}$$

Moreover let the function v be linear homogenous and strictly concave

$$v(x_k, a_k) = a_k \phi\left(\frac{x_k}{a_k}\right) \tag{4}$$

maximizing u with respect to $x_k$ implies then

$$0 = \phi'\left(\frac{x_k}{a_k}\right) - r_k w' \tag{5}$$

Concavity of v means that $\phi'$ is non-decreasing. Let f denote its inverse,

$$\frac{x_k}{a_k} = f(r_k \cdot w') \tag{6}$$

We shall treat $w'$, the marginal utility of time, as approximately constant in the range of trips considered. Trips $x_{ik}$ by all households $n_i$ at location i to destination k are then

$$x_{ik} = n_i a_k f(r_{ik} \cdot w') \tag{7}$$

where $r_{ik}$ is now the distance in time units, from i to k.

Formula (7) factors trip frequency into three components

- population $n_i$ at origin i
- attraction $a_k$ of destination k
- $f(w' r_{ik})$ the inhibiting effect of distance $r_{ik}$

conforming to the prototype of "gravity laws".

### Mathematical Note 1.5
### Alternative Derivation of an Exponential Distance Effect

In equation (1.4,5) $$\phi'\left(\frac{x}{a}\right) = w' r \tag{1}$$

(2)

set $w' = \mu$, and treat $w'$ as a constant and assume $\phi'$ to be logarithmic

$$\phi'\left(\frac{x}{a}\right) = -\ln\frac{x}{a} \tag{3}$$

so that

$$\phi(y) = -\int_o^y \ln t dt$$

$$= y(1 - \ln y) \qquad (4)$$

now, using (1.4.4)

$$v(x,a) = a\phi\left(\frac{x}{a}\right) = a \cdot \frac{x}{a}\left(1 - \ln\frac{x}{a}\right)$$

$$= (1 + \ln a)x - x\ln x \qquad (5)$$

A utility function

$$u = \sum_k v(x_{ik}, a_k) + \mu\left(T - \sum_k r_{ik}x_{ik}\right) = \sum_k (1 + \ln a_k)x_{ik} - \sum x_{ik}\ln x_{ik} + \mu\left(T - \sum r_{ik}x\right) \qquad (6)$$

maximized with respect to $x_{ik}$
will then generate trip demands

$$x_{ik} = a_k e^{-\mu r_{ik}} \qquad (7)$$

since

$$\frac{\partial u}{\partial x_{ik}} = 1 + \ln a_k - 1 - \ln x_{ik} - \mu r_{ik} = 0$$

implies

$$x_{ik} = a_k e^{-\mu r_{ik}}$$

It is interesting that an entropy term -x ln x emerges also in the likelihood maximization approach described in mathematical note 1.6.

**Mathematical Note 1.6**
**Maximum Likelihood and Entropy**

The following information is assumed known (Wilson 1969)

- number of trips originating in i, $A_i = \sum_k x_{ik}$
- number of trips terminating in k, $B_k = \sum_k x_{ik}$
- total mileage of travel $M = \sum_{ik} r_{ik} x_{ik}$

Given that all possible trips are equally likely otherwise, what is the most likely trip frequency $x_{ik}$ in the light of this information?
Write

$$x = \sum_{ik} x_{ik}$$

for the sum of all trips. The likely-hood function is them

$$L = \frac{x!}{\prod_{ik} x_{ik}!}$$

It may be approximated using Stirling's formula

$$L \approx \left(\frac{x}{e}\right)^x \Big/ \prod_{ik} \left(\frac{x_{ik}}{e}\right)$$

so that

$$\log L \approx x \ln x - \sum_{ik} x_{ik} \ln x_{ik}$$

Now $x = \sum_i A_i = \sum_k B_k$ is a given constant. Consider therefore $\max - \sum_{ik} x_{ik} \ln x_{ik}$ subject to the constraint

$$\sum_k x_{ik} = A_i$$

$$\sum_i x_{ik} = B_k$$

$$\sum_{ik} r_{ik} x_{ik} = M$$

a constrained entropy maximization.

Using Lagrange multipliers we have an unconstrained maximand

$$-\sum_{ik} x_{ik} \ln x_{ik} + \sum_i \lambda_i \left(-A_i + \sum_k x_{ik}\right)$$

$$+ \sum_k \mu_k \left(-B_k + \sum_i x_{ik}\right) + \upsilon \left(M - \sum_{i,k} r_{ik} x_{ik}\right)$$

maximization with respect to $x_{ik}$ yields then

$$-\ln x_{ik} - 1 + \lambda_i + \mu_k - \upsilon r_{ik} = 0$$

$$x_{ik} = e \cdot e^{\lambda_i} e^{\mu_k} e^{-\upsilon r_{ik}} \qquad \text{or}$$

$$x_{ik} = Q_i D_k e^{-\upsilon r_{ik}} \qquad \text{(say)}$$

This is a gravity law with exponential distance effect. Notice however the critical assumption that originations, terminations and total mileage must be given as data.

### Mathematical Note 1.7
### A Search Model

As a specific case consider that one has found a job and is now searching for acceptable accommodation near the job site. Let housing be available at a given constant density $\rho$, let the search begin at the job location and explore systematically an area up to distance r. The quantity of housing available within distance r depends also on the layout of the street system but will be proportional to $r^2$, say $\rho c r^2$.

Let Q(r) be the probability that none of this is acceptable. If the search is considered as a Poison process with success parameter $\lambda$ the probability of nothing suitable being found within distance r is

$$e^{-\lambda\rho cr^2} = e^{-\mu r^2} \qquad \text{(say)}$$

Then the probability of the search stopping at distance r is

$$-\frac{d}{dr}Q(r) = 2\mu r e^{-\mu r^2}$$
$$= 2\lambda c \cdot r\rho e^{-\mu r^2}$$

With employment N at the job site given one obtains a gravity model for home to work trips of length r to a given destination

$$x_i(r) = N2\lambda c\rho r e^{-\rho\lambda c r^2}$$

Here $c\rho r$ denotes potential origins at distance r , N the termination of trips and

$$e^{-\mu r^2}$$

the distance effect which in the context of a homogeneous two-dimensional environment of potential origins is an exponential function not of r but of $r^2$.

### Mathematical Note 1.8
### Shopping Trips, Demand, and Distance

The effect of distance on trade, i.e. sales, is usually considered as the result of transportation costs to be added to the product price at the supplier location – the "mill price"(Ch. 3). This applies to goods shipped between firms, but how are the costs of distance to be calculated for consumers who purchase goods making shopping trips?

Let q be the quantity of a good demanded for a given period T, and x the number of shopping trips during T. Why would more than one shopping trip be considered? Purchasing goods for later use ties up funds, incurs storage costs and possibly deterioration and also, in the case of true "shopping goods", a risk of obsolescence. Denote these holding costs by h. Total costs of purchase are then the sum of these holding costs per unit time plus the transportation cost incurred by x trips. The optimal number of trips per period T is then determined by

$$\underset{X}{\text{Min}}\, krx + h\frac{q}{x}T \qquad (1)$$

yielding

$$0 = kr - \frac{hqT}{x^2}$$
$$x = \sqrt{\frac{hqT}{kr}} \qquad (2)$$

so that the frequency of shopping varies as the square root of inverse distance. Costs incurred in a period T are then

$$k(q,r) = kr\sqrt{\frac{hqT}{kr}} + h\frac{qT}{\sqrt{\frac{hqT}{kr}}} = 2\sqrt{hkT}\cdot\sqrt{qr} \tag{3}$$

proportional to the square root of quantity demanded times distance, with a proportionality factor that is the geometric mean of the two cost f actors from storage and transportation. Since q is proportional to the period length T, the cost (3) is in fact proportional to T.

In the case of rectangular demand the cost to the consumer of one commodity unit adds up to

$$p = p_M + 2\sqrt{hkT}\sqrt{r} \tag{4}$$

(where $p_M$ is the mill price), so that the distance effect is proportional to the square root of distance.

In the case of linear demand the distance effect is given by a complicated expression in terms of square roots, since price including shopping costs also enters into the q of (3).

## References


Beckmann M, Persson O (1996) A Note on Citation and Distance. Institute for Futures Studies: Stockholm

Beckmann R (1994) On Knowledge Networks in Science: Collaboration Among Equals. Annals of Regional Science 28:233–244

Carrothers GAP (1956) An Historical Review of the Gravity and Potential Concepts in Human Interaction. Journal of the American Institute of Planners 22:94–102

Golob TF, Gustafson RL, Beckmann MJ (1973) An Economic Utility Approach to Spatial Interaction. Papers of the Regional Science Association 30:159–182

Hammer C, Ikle C (1957) Intercity Telephone and Airline Traffic Related to Distance and the "Propensity to Interact". Sociometrx 20,4:306–316

Leonardi G, Papageorgiou YY (1980) An Asymptotic Derivation of the Multinominal Logit Model. Sistemi Urbani 3:333–348

McFadden D (1974) Conditional Logit Analysis of Qualitative Choice Behavior. In Zarembka P (ed) Frontiers in Econometrics. Academic Press, New York, pp 105–142

Niedercom JH, Bechtold BV (1969) An Economic Derivation of the Law of Spatial Interaction. Journal of Regional Science 9:273–282

Ravenstein EG (1885) The Laws of Migration. Journal of the Royal Statistical Association 52:241–301

Stauffer SA (1940) Intervening Opportunities: A Theory Relating Mobility and Distance. American Sociological Review 5:845–867

Stewart JQ (1950) The Development of Social Physics. American Journal of Physics 5,2:239–253

Wartz W (1959) Towards a Geography of Price: A Study in Geo-Econometrics. University of Pennsylvania Press, Philadelphia, PA

Wilson A (1969) The Use of Entropy Maximizing Models. Journal of Transport Economics and Policy

CHAPTER 2

# Market Areas

The type of interaction at a distance that is of greatest interest to economists is that of exchange, i.e. buying and selling. Depending on the product, buyers may be found all over the world. A single sellers market is bounded only by the distances at which price plus transportation cost causes demand to reach zero or where, if the seller pays transportation cost, the price becomes equal to marginal production cost plus transportation cost.

When transportation rates are iso-tropic i.e. independent of direction, this determines a maximum market radius and a circular market area.

When several sellers offer a homogeneous product, buyers will choose the cheapest source. A seller is then the exclusive supplier in some territory, his *market area.*

When a seller has higher prices but lower transportation rates than his competitors, this market area begins at some distance from the seller's location. But otherwise the neighborhood of a seller is part of his market area and he enjoys a monopoly position there.

At the boundary between two market areas, prices plus transportation costs must be equal. When mill prices and transportation rates are equal, one buys from the nearest source; market areas of two firms are then separated by a straight line, the mean bisector of the line joining the two sellers.

When prices are different, the cheaper seller's market is bounded by a hyperbolic arc bending around the dearer rival.

When transportation rates are different, the supplier with the cheaper freight rate may compete with distant rivals and enclose their market areas with his own.

When the rates of transportation costs are the same for all and suppliers never farther from any buyer than the maximum market radius, then the entire region is disaggregated into a set of non-overlapping and mutually exclusive market areas.

It was observed by Launhardt (W. Launhardt 1885) that a lowering of freight rates results in an expansion of market areas of the firms with the lower production cost at the expense of those with higher production costs. A high cost firm loses its market entirely when undercut by the cheaper supplier at its own site.

With increasing returns to scale in production the firm that expands its production first then encroaches on and may eventually drive out its competing neighbors, even when all were of the same size to begin with. But as shown in chapter 7 coexistence of suppliers with increasing returns is possible in a stable spatial equilibrium. The stabilizing factor is of course transportation cost.

A three dimensional representation of prices for buyers in a circular (or unbounded) market area is known as *Launhardt's Cone* (Trichter). The loci of constant

**Fig. 2.1.** A Launhardt Cone

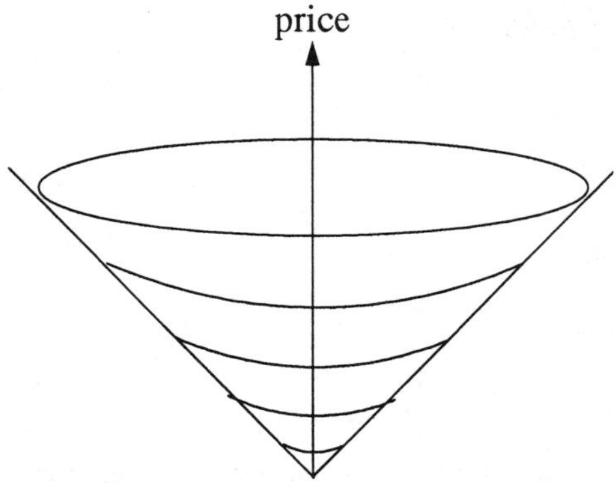


prices form a set of concentric circles. The boundary between two market areas can be found by intersecting these cones, which in the case of different prices and freight rates are placed at different levels and have different slopes (Figure 2.1).

Market areas exist not only for consumer goods but for anything that is traded, including industrial raw materials such as coal or oil.

When products are heterogeneous and prices comparable, it is not always advantageous to buy from the nearest supplier. In fact, with heterogeneous products or consumer preferences for particular suppliers, cross hauling is common, that is, buying from more distant suppliers according to preference. Hence a market area may enclose non-buyers as well. For homogenous commodities the market share of a seller falls from unity to zero as the boundary of the market area is crossed. For heterogeneous products market share, just as any other indicator of interaction, also falls with distance but not as abruptly (mathematical note 2.3). When market areas are defined to contain all points of positive sales, then they will overlap, when products are heterogeneous or buyers have preferences for other sellers.

A system of exhaustive and mutually exclusive market areas emerges, however, if we define each supplier's market area as that territory where his market share is dominant.

As a market boundary is crossed another supplier obtains the larger local market share. That market share should decrease with distance can be demonstrated by "random utility models" (mathematical note 1.3). Dominant market share defines market areas because market share decreases with distance.

In a dispersed industry, market areas of equal shape and size, bounded by line segments may emerge. These can be viewed as a tiling of the plane. There exist only three possible tilings with regular polygons: triangles, squares and hexagons. They approximate circles to an increasing degree. The economic advantage of a market area's shape is indicated by the average distance of its points from the center, considered as the location of a supplier. This average distance is proportional to total transportation cost when a unit (or constant amount) of a commodity is demanded everywhere and transportation cost is proportional to distance. Average distances have been calculated for polygonal market areas since Lösch (1940). Another indicator of economic (dis)advantage is the maximum distance also given in table 2.2 for the regular triangle, square, hexagon and the circle.

**Table 2.2.** Properties of regular polygons as functions of the radius R of the inscribed circle

| | *n* | *area (R)* | *average distance* $\bar{r}$ *(R)* | *maximum distance* |
|---|---|---|---|---|
| *Triangle* | 3 | $3 \tan \frac{\pi}{3} = 5.196\ R^2$ | .92 R | 2 R |
| *Square* | 4 | $4\ R^2$ | .765 R | 1.414 R |
| *Hexagon* | 6 | $6 \tan \frac{\pi}{6} = 3.46\ R^2$ | .702 R | 1.154 R |
| *Circle* | ∞ | $\pi = 3.14159\ R^2$ | $.\bar{6}$ R | R |

As expected, average and maximum distance fall as one approaches the circle as the ideal market area.

Supply areas are the exact counterpart to market areas with the parts of buyer and seller interchanged.

Consider next the special relations between trading centers considered as discrete and separate point locations. The manifest variables are now the links defined by commodity flows between such centers. Whether these links occur within areas that can be defined as market or supply areas is a question of flow configuration. Trade flows are generated by surpluses and deficits or excess supplies and excess demands at the given locations. In a spatial market equilibrium the flows accommodating these surpluses and deficits must constitute a commodity movement at minimum total transportation cost. Finding this minimum is the object of the "Transportation problem", the first linear program ever studied (Hitchcock 1941, Koopmans 1949).

The graph of cost minimizing flows can always be arranged as a tree, in which not more than a single path exists between any two points (flow directions ignored). A market area emerges where flows diverge from a particular point and a supply area where flows converge. Since neither need happen, a system of trading centers may not exhibit market or supply areas.

Consider also an extended spatial market with continuously distributed supply and demand. Actually it is excess demand (or its negative: excess supply) whose density need be defined over a certain region. An efficient i.e. transportation cost minimizing trade pattern requires that at any point (except a "singularity") there should be just one direction of flow, thereby defining a flow field in the region. Without singularities the flow direction varies continuously. Market areas emerge only when points exist from which flows issue in all directions (so-called "centers") and supply areas where flows converge from all directions (also called "centers").

A *market area* is recognizable by its pattern of commodity flows. The good moves (is shipped) from the supplier at a central location to the buyers throughout the market area along "shortest paths", i.e. cheapest routes. This same pattern of commodity movement from a apparent single source emerges also when suppliers and demanders are distributed continuously in space and form a competitive market (Beckmann 1952, Beckmann-Puu 1985).

The price will be lowest at the apparent source and increase by transportation cost along the paths of shipment. At a boundary separating two market areas, the direction of flow is reversed.

We note in conclusion that market area is a correlate of (potential) profits when and only when unit costs are constant and demand is rectangular, i.e. perfectly inelastic at a certain price, the "reservations price" (mathematical note 3.5.1).

Under uniform pricing and with uniform demand functions revenue is proportional to market area measured in terms of population rather than land units.

Under mill pricing and uniform demand functions, profits are proportional to market area in terms of population only when demand is inelastic.

### Mathematical Note 2.1
### The Shape of Market Areas

Let firm 1 with transportation costs $k = 1$ be located at $x = 0$, $y = 0$ and firm 2 with transportation costs $k < 1$ at $x = a$, $y = 0$ (figure 2.1a).

The boundary line is defined by

$$x^2 + y^2 = k(a-x)^2 + ky^2 \qquad (1)$$

or

$$(1-k)y^2 = ka^2 - 2akx + kx^2 - x^2$$

$$y^2 = \frac{k}{1-k}a^2 - 2a\frac{k}{1-k}x - x^2$$

$$y^2 + \left(x + \frac{ak}{1-k}\right)^2 = \frac{k}{1-k}a^2 + \frac{k^2}{(1-k)^2}a^2$$

This defines a circle with center at

$$x = -\frac{k}{1-k}a$$

Fig. 2.1. Circular market boundary between firms with same mill price and different transportation costs

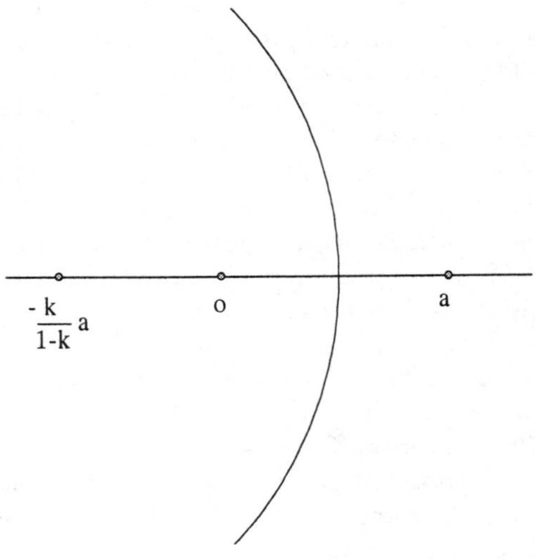

and radius

$$r = a\sqrt{\frac{k}{1-k} + \frac{k^2}{(1-k)^2}}$$

$$= \frac{ak}{1-k}\sqrt{1 + \frac{1-k}{k}}$$

$$= \frac{a\sqrt{k}}{1-k}$$

The point (a,0) lies outside the circle since

$$-\frac{k}{1-k}a + \frac{\sqrt{k}}{1-k}a < a$$

We have shown: when prices are equal but not transportation rates (cost per mile) the market area of the supplier with the higher transportation costs is a circle surrounded by the market area of the one with lower transportation costs. The center of this circle is on a line through the two locations away from the cheaper supplier. This transportation cost advantage is thus effective both at short and at long distances. When transportation costs are equal but prices are different, the boundary line becomes a hyperbola, defined as the locus of points whose distances from two given points differ by a constant. When both prices and freight rates differ, the separating line is defined by a quartic equation. The high price supplier will still be the cheaper one at large distances enclosing the other firm's market area with his own, provided freight rates are lower (Launhardt 1885). The market boundaries are best seen by intersecting the two firms' Launhardt cones.

Under a Manhattan metric the iso-vectures or loci of equal distance from a given origin (the supplier's location) are not circles but squares tilted at an angle of 45°, $\left(\frac{\pi}{2}\right)$ to the given road grid. Let the source point lie at the origin of the coordinate system. The iso-vectures are then defined by

$$|x| + |y| = \text{constant distance} = r$$

and form four linear segments at angles $\frac{\pi}{2}, \frac{3\pi}{2}, -\frac{\pi}{2}, -\frac{3\pi}{2}$

Thus, when prices are equal the boundaries between market areas are always straight lines at one of these four angles.

## Mathematical Note 2.2
## Market Areas as Regular Polygons

In comparing the effectiveness of geometric types of market areas the criterion of interest is average distance between the seller located at the center and the buyers assumed to be distributed at constant density throughout the market area. If $\rho$ (r) is density at distance r, then a constant density $\rho$ implies

$$\rho(r)dr = 2\pi\rho r dr \tag{1}$$

For a circular market area of radius R, the average distance is then

$$\bar{r}_0 = \frac{\int 2\pi\rho r^2 dr}{\int 2\pi\rho r dr} = \frac{\frac{1}{3}R^3}{\frac{1}{2}R^2} = \frac{2}{3}R \tag{2}$$

For a regular n-gon: triangle, square, pentagon, hexagon, one looks at a segment of $\frac{1}{2n}$ of its circumference (figure 2.2) with angle $\frac{\pi}{n}$

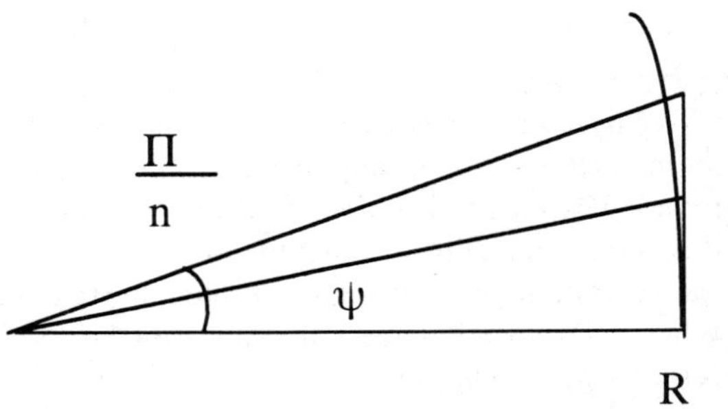


**Fig. 2.2.** Slice of $\frac{1}{2n}$ of an n-gon

Distance to the boundary along a ray at angle $\Psi$ is then $\frac{R}{\cos\psi}$ and average distance to all points in the triangle opened by angle $\frac{\pi}{n}$ is

$$\bar{r}_n = \frac{\int_0^{\frac{\pi}{n}} d\psi \int_0^{\frac{R}{\cos\psi}} r^2 dr}{\int_0^{\frac{\pi}{n}} d\psi \int_0^{\frac{R}{\cos\psi}} r dr}$$

$$= \frac{\frac{1}{3}\int_0^{\frac{\pi}{n}} d\psi \frac{R^3}{\cos^3\psi}}{\frac{1}{2}\int_0^{\frac{\pi}{n}} d\psi \frac{R^2}{\cos^2\psi}}$$

$$= h_n \cdot R$$

proportional to the radius R of the inscribed circle (figure 2.2) with

$$h_n = \frac{2}{3} \frac{\int_0^{\frac{\pi}{n}} \frac{d\psi}{\cos^3 \psi}}{\int_0^{\frac{\pi}{n}} \frac{d\psi}{\cos^2 \psi}} \tag{3}$$

$$= \frac{2}{3} \frac{\frac{1}{2}\left[\frac{\sin\frac{\pi}{n}}{\cos^2\frac{\pi}{n}} + \log\tan\frac{(n+2)\pi}{4n}\right]}{\tan\frac{\pi}{n}}$$

$$= \frac{1}{3} \frac{\left[\sin\frac{\pi}{n}\Big/\cos^2\frac{\pi}{n} + \log\tan\frac{(2+n)\pi}{4n}\right]}{\tan\frac{\pi}{n}}$$

The interesting cases n = 3, 4, 6 which permit a complete tessellation of the plane are listed in table 2.2 (cf also Puu 1997).

## Mathematical Note 2.3
## Models of Market Share

In the traditional literature of consumer research "Reilly's Law" is quoted asserting that market shares are inversely proportional to the square of distances. This is an example of a power law for market shares

$$\frac{m_i}{m_j} = \left(\frac{r_j}{r_i}\right)^{\alpha}$$

scaled so that market shares at equal distances are equal. (This need not be true when the two products are of unequal attractions). In a more general version the ratio of market shares, is proportional also to the ratio of sizes of the competing sellers (Beckmann 1969).

$$\frac{m_i}{m_j} = \frac{S_i}{S_j}\left(\frac{r_j}{r_i}\right)^{\alpha} \tag{1}$$

Since market shares must add up to one, the solution is

$$m_i = \frac{S_i r_i^{-\alpha}}{\sum_j s_j r_j^{-\alpha}} \tag{2}$$

An alternative postulate is that ratios of market shares depend on differences of distance

$$\frac{m_i}{m_j} = f\left(r_j - r_i\right) \tag{3}$$

where f is a monotone increasing function with f(0) = 1. We seek market shares as a function of distance that solve (2.1.3).

A solution satisfying $m_i$ (0) = 1, market share at the supplier's location is unity, is

$$m_j\left(r\right) = e^{-ar.} \tag{4}$$

yielding

$$\frac{m_i}{m_j} = e^{-a\left(r_i - r_j\right)} \tag{3a}$$

The exponential distance law (2.1.4) for market share can be derived from a "random utility model" (mathematical note 1.3).

## References


Beckmann M (1952) A Continuous Model of Transportation. Econometrica 20,4:643–660

Beckmann M, Puu T (1985) Spatial Economics: Density, Potential and Flow. North Holland: Amsterdam

Beckmann MJ (1968) Location Theory. Random House, New York

Beckmann M (1969) Market Share and Distance. The Swedish Journal of Economics, p 53–63

Hitchcock FL (1941) The Distribution of a Product from Several Sources To Numerous Qualities. Journal of Mathematics and Physics 20,3:224–230

Hyson CD, Hyson WP (1950) The Economic Law of Market Areas. Quarterly Journal of Economics 64:319–324

Koopman TC (1949) Optimum Utilization of the Transportation System. Econometrica 17 Supplement 136–146

Launhardt W (1885) Mathematische Begründung der Volkswirtschaftslehre. B.G. Teubner, Leipzig. English translation: Mathematical Principles of Economics (1993). Edward Elgar, Aldershot

Lösch A (1940) The Economics of Location. Yale University Press, New Haven, 1965, p 105–120

Puu T (1997/ Mathematical Location and Land Use Theory: An Introduction. Springer-Verlag, Heidelberg, New York

Reilly WJ (1931) The Law of Retail Gravitation. Knickerbocher Press, New York

CHAPTER 3

# Spatial Price Policy

A firm selling in spatially extended markets must chose prices for customers in different locations. Price policy thus means price setting, not price taking. The usual distinctions between market forms of monopoly, oligopoly, monopolistic and perfect competition must thus be observed. We begin with monopoly pricing. Competitive scenarios with price taking are studied in chapter 4. While the details of spatial pricing are of necessity mathematical, the main results of this chapter may be summarized verbally as follows. Three types of spatial pricing systems are widely used

- mill pricing or f.o.b.
- uniform delivered prices or c.i.f.
- discriminating prices set separately for each location, or for different zones.

Let per capita demand at all locations be subject to the same demand function but allow the number or density of customers to vary.

## Linear Demand Functions

The theory of spatial pricing is best developed for identical linear demand functions. To simplify the mathematics, the slope coefficient of this demand function may be standardized at unity by a suitable choice of the price unit

$$q = a - p \tag{1}$$

Initially we let the market radius, defining the size of the market area, be given. While market radius was the radius of the inscribed circle in comparing polygons (mathematical note 2.2) here in order to allow arbitrary shapes and density distributions, it is best to let market radius be the maximum distance from the supplier to any household in the market area. Shape of market area is defined by the vanishing of household densities.

To make comparisons between pricing systems, the price levels should first be equalized. This means that the uniform price should equal the mill price plus transportation cost over the average distance in the market area, that is, the price paid on average by customers served, where the average is taken with respect to locations rather than quantities sold.

When the (perfectly) discriminating price is set as the average of prices paid under mill and uniform pricing (see below), its average price level is then also the same.

Not surprising, at equal average price levels, the total quantities sold are the same for all three pricing systems and depend only on the average price level and on the radius of the market. More remarkable is the fact that then mill pricing and uniform pricing are equally profitable. The profit functions for mill and uniform pricing may in fact be transformed into each other when the uniform price equals the mill price plus average transportation cost.

Discriminatory pricing, even when just averaging the prices paid under mill and uniform pricing is more profitable in the same market area.

What are the profit maximizing prices? The best mill price turns out to be the average of demand intercept a and unit cost c minus half the transportation cost for the average distance

$$p_M = \frac{1}{2}\left(a + c - k\bar{r}\right) \tag{2}$$

and the optimal uniform price is half the average of demand intercept a and unit cost c plus half the transportation cost for the average distance $\bar{r}$

$$p_U = \frac{1}{2}\left(a + c + k\bar{r}\right) \tag{3}$$

The prices paid by buyers are then in fact the same on average, lending support to our previous assumption. Realized profits are then

$$G_M = G_U = \frac{1}{4}\left(a - c - k\bar{r}\right)^2 M \tag{4}$$

one fourth the population M of the market area times the square of the spread between demand intercept and average cost (including transportation).

For discriminatory pricing the extra profit equals

$$\frac{k^2}{4M}\left[\int r^2\rho(r)dr\int\rho(r)dr - \left(\int r\rho dr\right)^2\right] \tag{5}$$

proportional to the difference yielded by the Schwartz inequality.

When population densities are constant = ρ and markets circular so that

$$\bar{r} = \frac{2}{3}R \tag{6}$$

$$M = \pi\rho R^2 \tag{7}$$

maximum profits as a function of the market radius are

$$G_M(R) = G_U(R) = \frac{\pi\rho k^2}{4}\left(\frac{a-c}{k} - \frac{2}{3}R\right)^2 R^2 \tag{8}$$

For perfectly discriminating pricing they become

$$G_D(R) = \frac{\pi\rho k^2}{4}\left[\left(\frac{a-c}{k}\right)^2 \frac{R^2}{2} - \frac{2}{3}\frac{a-c}{k}R^3 + \frac{R^4}{4}\right] \tag{9}$$

The inequality (5) determines also the excess of transportation cost for uniform over mill pricing when quantities sold are the same. The welfare implication of this waste of transportation cost under uniform pricing are explored in mathematical note 15.7.

Transportation costs are wasted under uniform pricing since buyers demand the same quantity everywhere while under mill pricing quantity demanded falls off with distance.

Let now the market radius be chosen by the supplier. Maximizing profits means that the radius is expanded to the point where demand is zero under mill pricing and when marginal cost of production and delivery equals price under uniform pricing. Both turn out to be the same when profit maximizing prices are chosen.

With perfectly discriminatory pricing the market radius is chosen at the maximal economic level, that is when unit cost plus transportation cost reach the demand intercept

$$R_D = \frac{a-c}{k} \tag{10}$$

while the profit maximizing market radius under mill or uniform pricing is just $\frac{3}{4}$ of that.

$$R_M = R_U = \frac{3}{4}\frac{a-c}{k} \tag{11}$$

When the average distance under the optimal market radius is substituted in the price formula the profit maximizing uniform and mill prices become expressions in terms of demand and cost coefficients only. Simple expressions result for uniform population densities and circular market areas

$$p_M = \frac{a}{4} + \frac{3}{4}c \tag{12}$$

$$p_U = \frac{3}{4}a + \frac{c}{4} \tag{13}$$

They are weighted averages of demand intercept and unit cost, the weights of $\frac{3}{4}$ and $\frac{1}{4}$ applied to yield the higher uniform and the lower mill price, respectively.

The profit maximizing discriminatory price does not depend on average distance and is hence the same under all choices of market radii.

Achieved profits for the optimal market radius and homogeneous population density are now for mill and uniform pricing, using in (4)

$$\begin{aligned} G_k^* &= \frac{\pi\rho 9k^2}{256}\left(\frac{a-c}{k}\right)^4 \\ &\doteq 0.110\rho\frac{(a-c)^4}{k^2} \end{aligned} \tag{14}$$

while discriminatory pricing yields using (10) in (9)

$$G_D^* = \frac{\pi\rho k^2}{24}\left(\frac{a-c}{k}\right)^4$$

$$\doteq 0.131\rho\frac{(a-c)^4}{k^2} \qquad (15)$$

Their ratio is

$$\frac{G_D}{G_M} = \frac{32}{27} \doteq 1.185$$

Finally let demand functions be linear, but not identical.

The slope coefficient b(r) will then affect the difference between uniform and mill price for which sales become equal. It is now

$$p_U - p_M = k\frac{\overline{br}}{\overline{b}}$$

where the averages are weighted with population density as before

$$\overline{b} = \frac{\int b(r)\rho(r)dr}{\int \rho(r)dr}$$

$$\overline{br} = \frac{\int b(r)r\rho(r)dr}{\int \rho(r)dr}$$

but it does not change relative profitabilities.

A negative correlation between intercepts a(r) and distance makes uniform pricing more profitable. This situation may occur for the following reasons: delivery is delayed, the product is less well known, it is less suited to local tastes or (for newsworthy items) it is less relevant (Beckmann 1989).

## Nonlinear Demand

Less is known for nonlinear demand, when market shapes and population densities are arbitrary.

The fact that transportation cost is wasted under uniform pricing compared to mill pricing remains true for all demand functions for which quantity demanded decreases with price. (Hence, it is not true for rectangular demand if the market radius chosen is the same – which it is not under profit maximizing choices (cf mathematical note 3.5.1).

For convex and concave identical demand functions comparisons are possible with regard to quantities sold and profits, when the price levels are once more equalized (uniform price = mill price plus average transportation cost paid, averaged over distance – not units sold).

For concave demand functions, uniform pricing yields the greater sales and profits, and for convex demand functions mill pricing does. With profit maximizing prices – dropping the assumption of equal price levels – the comparison is still valid for maximal profits: greater for uniform price when demand is concave and lower when demand is convex.

Since linear functions are both convex and concave, the equality of quantities and profits under mill and uniform pricing is consistent with this.

When population densities are constant, sales, profits, and profit maximizing prices and market sizes as well as realized profits, can be calculated readily for some standard demand functions: rectangular (concave) and exponential (convex) as well as constant elasticity demand functions (convex) (mathematical note 3.5).

## Cartels and Collusive Oligopoly

So far we have considered price policies under monopoly. While firms may be considered monopolists within their respective market area, the extent of that market area will itself depend upon price. True monopoly is realized only, when market areas have been fixed through allocation under a collusive agreement, a cartel, or when firms are set sufficiently far apart so that market areas chosen under their profit maximizing price policy do not overlap.

In the first case the market radius R defined as the maximum distance to customers served, is given, in the second case it can still be chosen.

Rivalry with neighboring firms over market areas falls under non-collusive spatial oligopoly, to be considered in chapter 7. Depending on distance and the price policy chosen, Cournot oligopoly i.e. mutual adjustment is possible for mill and discriminatory pricing (mathematical note 7.4), while Bertrand oligopoly is a natural consequence of uniform pricing (mathematical note 7.3).

The best known example of collusive price setting in an oligopoly is the Basing Point System, also known as Pittsburgh Plus. Here firms agree to charge c.i.f. prices (that include transportation cost) as if delivery were made from the basing point. The intention is, of course, to protect firms located at the basing point from being undercut by any one located elsewhere. Basing point pricing has also been used in the cement industry. This system, illegal now everywhere, is of merely historical interest.

Collusive pricing may be instituted in various other ways.

- by allocating exclusive market areas thus permitting each firm to act as a monopolist
- by an agreement on minimum prices (mill or uniform) presumably the profit maximizing price of a price leader or a dominant firm
- through a cartel both fixing the price and assigning sales territories (market areas) to its members, not always in a cost minimizing way.

### Mathematical Note 3.1
### Sales and Average Prices

A comparison of pricing systems is most revealing when demand functions are of the simplest type, linear and identical for all customers, while the number or density of buyers is allowed to depend on location. Cost functions must be specified when comparisons are made with respect to profits and welfare. To begin with we assume a given market radius R, consumer density $\rho(r)$ and consider sales, and average prices per customer.

For uniform density $\rho$ customers in a ring of width dr at distance r are

$$\rho(r)dr = 2\pi\rho dr$$

Let $p_M$ be the mill price, $p_U$ the uniform price.

*Sales:*

$$\text{uniform pricing} \qquad f(p_U)\int_0^R \rho(r)dr \tag{1}$$

$$\text{mill pricing} \qquad \int_0^R f(p_M + kr)\rho(r)dr \tag{2}$$

Assume average price to be the same under each system

$$p_U = \overline{p}_M = \frac{\int (p_M + kr)\rho dr}{\int \rho dr} = p_M + k\overline{r}$$

where

$$\overline{r} = \frac{\int_0^R r\rho(r)dr}{\int_0^R \rho(r)dr} \tag{3}$$

is the average distance to customers.

Recall from chapter 2 that in a one-dimensional market of radius R with uniform density

$\rho(r) = \rho$ and average distance is

$$\overline{r}(R) = \frac{R}{2}$$

In a two-dimensional market with uniform people density

$$\rho(r)dr = 2\pi r dr$$

or with Manhattan metric

$$\rho(r)dr = 4rdr$$

the average distance is

$$\overline{r}(R) = \frac{2}{3}R$$

Now a well-known property of convex and concave functions is

$$f(p_M + k\bar{r}) \lesseqgtr \frac{\int f(p_M + kr)\rho dr}{\int \rho(r)dr} \tag{4}$$

according as f (p) is $\begin{Bmatrix} \text{convex} \\ \text{linear} \\ \text{concave} \end{Bmatrix}$

Comparing (1) and (2) in the light of (4), we see:
When the uniform price is set equal to the mill price plus average transportation cost (weighted by the density of customers) so that average prices are the same then under uniform pricing average sales are $\begin{Bmatrix} \text{less} \\ \text{equal} \\ \text{greater} \end{Bmatrix}$ compared to mill pricing depending on whether quantity demanded is a $\begin{Bmatrix} \text{convex} \\ \text{linear} \\ \text{concave} \end{Bmatrix}$ function of price.

Since populations are the same, the comparison is valid also for total sales (aggregate demand).

Thus, for identical linear demand functions mill or uniform pricing makes no difference to total sales. Of course, under mill pricing more is sold near the firm, in fact, up to the average distance and less beyond average distance, which should result in less transportation cost, which we will now examine.

**Mathematical Note 3.2**
**Transportation Cost**

For arbitrary identical demand functions f(p) consider a mill price $p_M$ and a uniform price set

$$p_U = p_M + k\hat{r}$$

to generate equal sales
(For linear demand $\hat{r} = \bar{r}$ ).

$$\int_0^R f(p_M + k\hat{r})\rho dr = \int_0^R f(p_M + kr)\rho dr \tag{5}$$

The mean value theorem of the calculus assures that

$$0 \le \hat{r} \le R$$

By definition

$$0 = \int_0^R [f(p_M + kr)\rho dr - f(p_M + k\hat{r}]\rho dr \tag{3}$$

Now compare transportation cost

$$T_U = f(p_U + k\hat{r})\int_0^R r\rho dr$$

$$T_M = \int_0^R f(p_M + kr) r\rho dr$$

Using (5)

$$T_M - T_U = k\int_0^R [f(p_M + kr) - f(p_M + k\hat{r})](r - \hat{r})\rho dr \tag{6}$$

$$< 0$$

since downward sloping demand functions mean that the bracket and the parenthesis under the integral have opposite signs.

Thus for equal sales, transportation costs under mill pricing are less than under uniform pricing, or transportation per sale is cheaper under mill pricing than under uniform pricing. This proof may be extended to any transportation cost function k(r) that is increasing with distance r.

From now on we assume constant unit production cost which has been customary in location theory. A fixed cost F is also assumed but will not appear explicitly in this chapter.

**Mathematical Note 3.3**
**Profits**

These are: for a given

uniform price $$G_U = f(p_U) \int_0^R [p_U - c - kr]\rho(r)dr \tag{7}$$

and given mill price

$$G_M = (p_M - c) \int_0^R f(p_M + kr)\, \rho(r)dr \tag{8}$$

and under discriminatory pricing, when p(r) is set separately for each r

$$\int [p(r) - c - kr] f(p(r))\rho(r)dr \tag{9}$$

Assume once more a uniform price equal to the mill price plus average transportation cost

$$p_U = p_M + k\bar{r} \text{ with } \int_0^R (r - \bar{r})\rho dr = 0.$$

Now

$$G_M - G_U = \int_0^R f(p_M + kr)[p_M - c]\rho dr$$

$$- f(p_M + k\bar{r})\int_0^R [p_M - c + k(\bar{r} - r)]\rho dr$$

$$= (p_M - c) \int_0^R [f(p_M + kr) - f(p_M + k\bar{r})]\rho dr$$

$$\gtreqless 0 \text{ according as f is } \begin{Bmatrix} \text{convex} \\ \text{linear} \\ \text{concave} \end{Bmatrix}$$

For concave demand, profits can be further increased by choosing a uniform price not necessarily equal to the mill price (optimal for that pricing strategy) plus average transportation cost. Profitability for convex and concave demand was first compared by Stevens and Rydell (1966).

From now on, in determining prices and profits, we consider only linear demand functions. The demand function is

$$f(p) = a - p \tag{10}$$

where we have standardized quantity units to obtain slopes of unity.

Consider first mill and uniform pricing with profit functions

$$G_M = (p_M - c)\int_0^R [a - p_M - kr]\rho dr - F \tag{11}$$

$$G_U = (a - p_U)\int_0^R [p_U - c - kr]\rho dr - F \tag{12}$$

Suppose we let

$$p_M - c = a - p_U \quad \text{or} \quad p_M + p_U = a + c \tag{13}$$

Then expressions (11) and (12) for profits become identical. The profit maximum has therefore the same value under mill or uniform pricing (Beckmann and Ingene 1976). The profit maximizing prices are then obtained from each other by the transformations

$$p_U = a + c - p_M \tag{13}$$

or

$$p_M = a + c - p_U$$

The profit maximizing *mill price* is determined by

$$\max_{p_M} \; [p_M - c]\int [a - p_M - kr]\,\rho dr$$

yielding

$$0 = \int \left[c - p_M + a - p_M - kr\right]\rho dr$$

$$p_M = \frac{a+c}{2} - \frac{k}{2}\bar{r} \qquad (14)$$

The profit maximization *uniform price* is found by

$$\max \left(a - p_U\right) \int_0^R \left[p_U - c - kr\right]\rho dr$$

or

$$0 = \int_0^R \left[-p_U + c + kr + a - p_U\right]\rho dr$$

$$p_U = \frac{a+c}{2} + \frac{k\bar{r}}{2} \qquad (15)$$

A comparison of (14) and (15) shows that for linear identical demand, the profit maximizing uniform and mill prices will differ by average transportation cost

$$p_U = p_M + k\bar{r} \qquad (16)$$

The two relations

$$p_M + p_U = a + c \qquad (13)$$

and

$$p_U = p_M + k\bar{r} \qquad (16)$$

together will also determine the profit maximizing mill and uniform prices.

Consider also the profit maximizing prices under perfect discrimination. Differentiating (9) with respect to p(r) under the integral yields, for f(p) = a-p

$$0 = \int \left(\left[a - p(r)\right] - \left[p(r) - c - kr\right]\right)\rho(r)dr$$

or

$$p(r) = \frac{1}{2}\left(a + c + kr\right) \qquad (17)$$

(Singer 1937).

Comparing (11) and (14), (15) shows that the perfectly discriminating price averages prices paid under mill and uniform pricing

$$p_D(r) = \frac{1}{2}\left(p_U + p_M + kr\right) \qquad (18)$$

Prices paid as a function of distance are shown in figure 3.1. They are equal at the average distance $\bar{r}$.

**Fig. 3.3.** Prices paid by customers

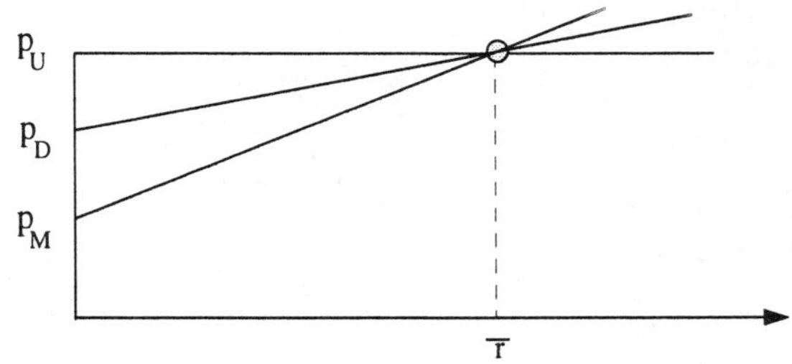


The achieved profits are now

$$G_M(R) = (p_M - c)\int_0^R [a - p_M - kr]\rho dr$$

$$= \frac{a - c - k\overline{r}}{2} \cdot \int \left[\frac{a - c + k\overline{r}}{2} - kr\right]\rho(r)dr$$

$$= \frac{k^2}{4}\left[\frac{a-c}{k} - \overline{r}(R)\right]^2 M(R) \tag{19}$$

where M(R) is total population in the market area of radius R

$$M(R) = \int_0^R \rho(r)dr$$

$$G_U = [a - p_U]\int_0^R [p_U - c - kr]\rho dr$$

$$= \left[\frac{a - c - k\overline{r}}{2}\right]\int_0^R \left[\frac{a - c + k\overline{r}}{2} - kr\right]\rho(r)dr$$

$$= \frac{k^2}{4}\left[\frac{a-c}{k} - \overline{r}(R)\right]^2 M(R) \tag{19}$$

For constant population density this becomes

$$G_M(R) = G_U(R) = \frac{k^2\pi\rho}{4}\left[\frac{a-c}{k} - \frac{2}{3}R\right]^2 R^2 \tag{20}$$

Profits per customers served are thus $\frac{1}{4}$ the square of the spread between demand intercept a and unit cost c plus transportation cost for the average distance.

Profits increase with market radius R for

$$0 \le R \le \frac{3}{4}\frac{a-c}{k}$$

since

$$\frac{dG_M}{dR} \sim -\frac{4}{3}R^2\left(\frac{a-c}{k}-\frac{2}{3}R\right)$$

$$+2R\left(\frac{a-c}{k}-\frac{2}{3}R\right)^2$$

$$\sim -\frac{4}{3}R+2\left(\frac{a-c}{k}-\frac{2}{3}R\right)$$

$$\sim -\frac{4}{3}R+\frac{a-c}{k}$$

For discriminatory pricing, achieved profits are

$$G_D(R)=\int_0^R\left[\frac{a-c-kr}{2}\right]^2\rho(r)dr$$

$$=\frac{k^2}{4}\int_0^R\left[\frac{a-c}{k}-r\right]^2\rho(r)dr$$

$$=\frac{k^2}{4}\left[\left(\frac{a-c}{k}\right)^2-2\left(\frac{a-c}{k}\right)\overline{r}+\overline{r^2}\right]M$$

For homogeneous population density and circular markets this simplifies to

$$G_D(R)=\pi\rho k^2\left[\left(\frac{a-c}{k}\right)^2\frac{R^2}{\overline{4}}-\frac{a-c}{k}\frac{R^3}{\overline{3}}+\frac{R^4}{\overline{8}}\right] \tag{21}$$

Profits increases with market radius R for

$$R\le\frac{a-c}{k}$$

The dimension of $\rho$ being $\frac{1}{\text{price}\cdot\text{area}}$, the dimension of profits in (20) and (21) is price. (Area is integrated out).

Observe that profits depend only on the standardized variable $\frac{a-c}{k}$ k, and market radius R.

When the demand parameters a and b in a linear demand function

$$q=a-bp$$

are not uniform but depend on location

$$a=a(r) \qquad b=b(r)$$

quantities sold and profits are no longer identical under mill and uniform pricing with

$$p_U = p_M + k\bar{r}$$

Rather the following is true. With demand functions

$$q = a(r) - b(r)p$$

sales are now under mill pricing

$$Q_M = \int \left[a(r) - b(r)p_M - b(r)kr\right]\rho(r)dr$$

$$= \left[\bar{a} - \bar{b}p_M - k\overline{br}\right]M \tag{22}$$

and under uniform pricing

$$Q_U = \int_0^R \left[a(r) - b(r)p_U\right]\rho(r)dr$$

$$= \left[\bar{a} - \bar{b}p_U\right]M \tag{23}$$

Sales are now equalized when

$$p_U = p_M + k\frac{\overline{br}}{\bar{b}} \tag{24}$$

where

$$\overline{br} = \frac{\int b(r)r\rho(r)dr}{\int b(r)\rho(r)dr}$$

and

$$\bar{b} = \frac{\int b(r)\rho(r)dr}{\rho(r)dr}$$

The profit maximizing prices are easily found to be

$$p_M = \frac{1}{2}\left(\frac{\bar{a}}{\bar{b}} + c - k\frac{\overline{br}}{\bar{b}}\right) \tag{25}$$

and

$$p_U = \frac{1}{2}\left(\frac{\bar{a}}{\bar{b}} + c + k\frac{\overline{br}}{\bar{b}}\right) \tag{26}$$

and

$$p_U - p_M = k\frac{\overline{br}}{\overline{b}}$$

equalizing sales.

Comparing profits

$$G_M = \frac{1}{2}\left(\frac{\overline{a}}{\overline{b}} - c - k\frac{\overline{br}}{\overline{b}}\right)\int_0^R \left[a - \frac{b}{2}\left(\frac{\overline{a}}{\overline{b}} + c - k\frac{\overline{br}}{\overline{b} - 2kr}\right)\right]\rho(r)dr$$

$$= \frac{1}{4}\left(\frac{\overline{a}}{\overline{b}} - c - k\frac{\overline{br}}{\overline{b}}\right)\left[\overline{a} - \overline{br} + k\overline{br} - 2k\overline{br}\right]M$$

$$= \frac{1}{4\overline{b}}\left(\overline{a} - \overline{b}c - k\overline{br}\right)^2 M \tag{27}$$

$$G_U = \int_0^R \left[a - \frac{b}{2}\left(\frac{\overline{a}}{\overline{b}} + c + k\frac{\overline{br}}{\overline{b}}\right)\right]\frac{1}{2}\left[\frac{\overline{a}}{\overline{b}} - c + k\frac{\overline{br}}{\overline{b}} - 2kr\right]\rho(r)dr$$

$$= \frac{1}{4\overline{b}}\left[\overline{a} - \overline{b}c - k\overline{br}\right]^2 M - \frac{k^2}{4}\left[\overline{ar} - \overline{a}\,\overline{r}\right]M \tag{28}$$

A varying slope coefficient b(r) does not change relative profitability, but the intercept does. A negative correlation

$$\overline{ar} < \overline{a}\,\overline{r} \tag{29}$$

between intercept and distance makes uniform pricing more profitable.

## Mathematical Note 3.4
## Market Radius

Drop now the assumption of a given market radius and allow markets to be extended at liberty. Consider first the profit maximizing radius for given prices

*Mill pricing:*

$$\underset{R}{\text{Max}}\left(p_M - c\right)\int_0^R f\left(p_M + kr\right)\rho dr$$

$$0 = \left(p_M - c\right)f\left(p_M + kR\right)\rho\left(R\right)$$

which for positive profit margins $p_M - c$ implies

$$\rho\left(R\right) f\left(p_M + kR\right) = 0 \tag{1}$$

The market radius is expanded to the distance R where quantity demanded at prices paid by buyers fall to zero, or to where no customers exist.

*Uniform pricing*

$$\underset{R}{\text{Max}}\; f(p_U)\int_0^R [p_U - c - kr]\rho dr$$

$$0 = f(p_U)\cdot[p_U - c - kR]\rho(R) = 0 \tag{2}$$

The market radius R is now the distance at which marginal production plus transportation cost just equals the uniform price or customer density has fallen to zero.

*Discriminatory pricing*

$$\underset{R}{\text{Max}} \int_0^R [p(r) - c - kr]\, f(p(r))\rho dr$$

$$0 = [p(R) - c - kR] f(p(R))\rho(R) \tag{3}$$

But the profit maximizing price is found by the condition

$$p - c - kr = -\frac{f}{f'} > 0 \tag{4}$$

which shows positive profits everywhere. Therefore sales stop at a point where either the quantity demanded is zero or the customer density has fallen to zero.

So far, prices were assumed given, but they must be newly set for the market radius chosen, (except for perfect discrimination), that is prices and market radius must be chosen simultaneously for maximal profits.

*Mill pricing*

$$\underset{p,R}{\text{Max}}\,(p - c)\int_0^R [a - p - kr]\rho dr$$

yielding

$$0 = \int [a - p - kr - p + c]\rho dr$$

$$0 = \int [a - c - kr - 2p]\rho dr$$

and

$$(p - c)[a - p - kR] = 0$$

so that

$$p = \frac{a + c - k\bar{r}}{2} \qquad a - p - kR_M = 0$$

For uniform population density and circular markets

$$\bar{r} = \frac{2}{3}R$$

so that

$$\frac{a-c+\frac{2}{3}kR_m}{2}-kR_M=0$$

$$\frac{2}{3}kR=\frac{a-c}{2}$$

$$R_M=\frac{3}{4}\frac{a-c}{k} \tag{5}$$

Notice that $\frac{a-c}{k}$ is the maximal market radius, attainable under marginal cost or under perfectly discriminatory cost pricing (see below).

*Uniform pricing*

$$p_U=\frac{a+c}{2}+\frac{k\,\bar{r}}{2}=c+kR_U \qquad \bar{r}=\frac{2}{3}R_U$$

$$\frac{a-c}{2}=k\left(R_U-\frac{R_U}{3}\right)$$

$$R_U=\frac{3}{4}\frac{a-c}{k}$$

the same as for mill pricing

$$R_U=R_M \tag{6}$$

For *perfectly discriminating pricing*, the maximal attainable market radius is in fact realized

$$0=\left[a-\frac{a+c}{2}-k\frac{R}{2}\right]\left[\frac{a+c}{2}+k\frac{R}{2}-c-kR\right]$$

$$=\frac{1}{4}(a-c-kR)^2$$

so that

$$R_D=\frac{a-c}{k} \tag{7}$$

We can now calculate the profit maxima for the profit maximizing market radii.

$$G_M(R)=\frac{k^2}{4}\left[\frac{a-c}{k}-\bar{r}\right]^2 M$$

Substituting

$$R=\frac{3}{4}\frac{a-c}{k} \quad \bar{r}=\frac{2}{3}R \quad M=\pi\rho R^2 \tag{8}$$

in (3.3.19)

$$\hat{G}_M = \frac{k^2}{4}\pi\rho \cdot \frac{1}{4}\left(\frac{a-c}{k}\right)^2 \cdot \left(\frac{2}{4}\right)^2\left(\frac{a-c}{k}\right)^2$$

$$= \frac{9}{256}\pi\rho\frac{(a-c)^4}{k^2} \tag{9}$$

$$\approx 0.0110\rho\frac{(a-c)^4}{k^2}$$

For perfect discrimination

$$\hat{G}_D = \frac{k^2\pi}{2}\int_0^{\frac{a-c}{k}}\left[\frac{a-c}{k}-r\right]^2 rdr \tag{10}$$

$$= \frac{k^2\pi\rho}{2}\left(\frac{a-c}{k}\right)^4\left[\frac{1}{2}-\frac{2}{3}+\frac{1}{4}\right]$$

$$= \frac{k^2}{24}\left(\frac{a-c}{k}\right)^2 M-F$$

$$= \frac{\pi\rho}{24}\frac{(a-c)^4}{k^2}$$

Notice that

$$\frac{1}{24} > \frac{9}{256}$$

confirming that price discrimination is more profitable than either uniform or mill pricing.

The choices $\frac{3}{4}\frac{a-c}{k}$ or $\frac{a-c}{k}$ of a market radius have interesting implications for transportation demand as a function of its price. It can be shown that transportation volume is a cubic and transportation mileage a quartic function of $\frac{1}{k}$, meaning that their elasticities of demand are, respectively, 3 and 4. The derivation is straight forward and is left to the reader.

Launhardt was aware of the enormous increase in transportation demanded as the result of the reduction to one tenth of freight rates for over land transportation cost by the railroads (Launhardt 1885)

## 3.5 Examples

### 3.5.1 *Rectangular Demand Curve*

**Fig. 3.5.1.** Rectangular Demand Curve

$$f(p) = \begin{cases} 0 & p > a \\ 1 & p \le a \end{cases} \tag{1}$$

This demand function is concave. The profit maximizing uniform price is clearly

$$p_U = a$$

yielding profits

$$G_U = \rho \int_0^R [a - c - kr]\, 2\pi r dr = \pi\rho \left[ (a-c)R^2 - \frac{2}{3}kR^3 \right] \tag{2}$$

The optimal sales radius R is determined by

$$a - c - kR = 0 \qquad R_U = \frac{a-c}{k} \tag{3}$$

so that

$$G_U = \pi \frac{\pi\rho}{3} \frac{(a-c)^3}{k^2} \tag{4}$$

Sales are then

$$Q_U = \pi\rho R^2 = \pi\rho \left( \frac{a-c}{k} \right)^2 \tag{5}$$

The consumers' surplus is zero. Welfare thus equals profits

$$W_U = G_U = \frac{\pi\rho}{3} \frac{(a-c)^3}{k^2} \tag{6}$$

Under mill pricing

$$G_M = (p-c)\rho\pi R^2$$

where R is determined by

$$p + kR = a \qquad R_M = \frac{a-p}{k} \tag{7}$$

For $p_M > c$ it follows that $R_M < R_U$
Now

$$G_M = \max_p G_M(R) = \max_p (p-c)\pi\rho\left(\frac{a-p}{k}\right)^2$$

yields

$$0 = (a-p)^2 - 2(a-p)(p-c)$$

$$a - p = 2\,(p-c)$$

$$p_M = \frac{a}{3} + \frac{2}{3}c \tag{9}$$

Hence the optimal market radius is, using (7),

$$R_M = \frac{2}{3}\frac{a-c}{k} \tag{10}$$

$$G_M = \pi\rho\frac{a-c}{3}\cdot\left[\frac{2}{3}(a-c)\right]^2$$

$$= \frac{4}{27}\pi\rho\frac{(a-c)^3}{k^2} \tag{11}$$

Thus

$$\frac{G_M}{G_U} = \frac{4}{9} \tag{12}$$

Since the rectangular demand function is an extreme among concave functions, this ratio is presumably the lowest possible for profits under the two types of pricing systems.

The ratio of market radii is $\dfrac{R_M}{R_U} = \dfrac{2}{3}$

The ratio of sales is therefore

$$\frac{Q_M}{Q_U} = \left(\frac{R_M}{R_U}\right)^2 = \frac{4}{9} \tag{13}$$

the same as the ratio of profits.

Welfare under mill pricing is the sum of consumers' surplus

$$S_M = 2\pi\rho\int_0^R (a-p-kr)r\,dr = 2\pi\rho\left[(a-p)\frac{R^2}{2} - k\frac{R^3}{3}\right]$$

$$= \pi\rho\frac{(a-c)^3}{k^2}\cdot\left[\left(\frac{2}{3}\right)^3 - \left(\frac{2}{3}\right)^4\right] \tag{14}$$

$$= \frac{8}{81}\pi\rho\frac{(a-c)}{k^2}$$

and profits

$$\frac{4}{27}\pi\rho\frac{(a-c)^3}{k^2} \tag{11}$$

$$W_M = \frac{20}{81}\pi\rho\frac{(a-c)^2}{k^2} < \frac{1}{3}\pi\rho\frac{(a-c)^3}{k^2} = W_U \tag{15}$$

Under uniform pricing welfare, accruing only to firms, is by $\frac{7}{81}$ larger than welfare under mill pricing. Since just one unit is demanded at every distance, mill pricing induces no savings in transportation cost here.

### 3.5.2 *Exponential Demand Function*

An exponential demand function

$$q(p) = e^{-p} \tag{16}$$

is convex.

Under uniform pricing

$$G_U = e^{-p}\int_0^R (p-c-kr)2\pi\rho r dr$$

with

$$R = \frac{p-c}{k} \tag{17}$$

profits are

$$G_U = \frac{\pi\rho}{3}e^{-p}\frac{(p-c)^3}{k^2} \tag{18}$$

and the most profitable price is determined by

$$0 = -e^{-p}(p-c)^3 + 3(p-c)^2 e^{-p}$$

$$p_U = c + 3 \tag{19}$$

a mark up of 3.

For mill pricing the market radius is infinite. Profits are

$$G_M = (p-c)\int_0^{\infty} e^{-p-kr} 2\pi\rho r dr$$

$$= 2\pi\rho e^{-p}(p-c)\int_0^{R} e^{-kr} r dr$$

$$= \frac{2\pi\rho}{k^2} e^{-p}(p-c) \tag{20}$$

The profit maximizing price is then given by

$$0 = \frac{d}{dp} e^{-p}(p-c) = -e^{-p}(p-c) + e^{-p}$$

$$p_M = c + 1 \tag{21}$$

a mark up of only 1,
yielding profits

$$\hat{G}_M = \frac{2\pi\rho}{k^2} e^{-c-1} \tag{22}$$

while maximal profits for uniform pricing are

$$\hat{G}_U = \frac{\pi\rho}{3} e^{-c-3} \cdot \frac{27}{k^2} = 9\frac{\pi\rho}{k^2} e^{-c-3} \tag{23}$$

The profit ratio is

$$\frac{\hat{G}_M}{\hat{G}_U} = \frac{2e^{-1}}{9e^{-3}} = \frac{2}{9} e^2 = 1.642 \tag{24}$$

confirming the greater profitability of mill pricing when demand functions are convex.

### 3.5.3
### *Constant Elasticity Demand Function*

A constant elasticity of demand $\varepsilon$ means a demand function

$$q = p^{-\varepsilon} \tag{25}$$

With uniform pricing profits are

$$G_U = p^{-\varepsilon}\int (p-c-kr) 2\pi\rho r dr$$

The profitable sales radius is, as always,

$$R_U = \frac{p-c}{k} \tag{26}$$

yielding profits

$$G_U = \frac{\pi\rho}{3k^2} p^{-\varepsilon}(p-c)^3$$

The profit maximizing price is the then determined by

$$0 = \frac{\partial}{\partial p} p^{-\varepsilon}(p-c)^3$$

$$= -\varepsilon p^{-\varepsilon-1}(p-c)^3 + 3p^{-3}(p-c)^2$$

from which

$$p_U = \frac{\varepsilon}{\varepsilon - 3} c \tag{27}$$

This profit maximum and associated price exist only when $\varepsilon > 3$, a rather stringent requirement for elasticity. The market radius is then

$$R_U = \frac{3}{\varepsilon - 3} \cdot \frac{c}{k} \tag{28}$$

and sales are

$$S_U = \left(\frac{\varepsilon}{\varepsilon - 3} c\right)^{-\varepsilon} \cdot \frac{\pi\rho}{k^2} \left(\frac{3c}{\varepsilon - 3}\right)^2$$

$$= \frac{9\pi\rho \cdot (\varepsilon - 3)^{\varepsilon - 2} \varepsilon^{-\varepsilon}}{k^2 c^{\varepsilon - 2}} \tag{29}$$

and profits

$$G_U = \frac{\pi\rho}{3k^2} \cdot \left(\frac{\varepsilon - 3}{c}\right)^{\varepsilon - 3} 3^3 \varepsilon^{-\varepsilon}$$

$$= 9\frac{\pi\rho}{k^2} \varepsilon^{-\varepsilon} (\varepsilon - 3)^{\varepsilon - 3} c^{3-\varepsilon} \tag{30}$$

a decreasing function of k and c.

Consider now mill pricing.

The market radius is unbounded. Profits are

$$G_M = (p-c)2\pi\rho\int_0^\infty (p+kr)^{-\varepsilon} r\,dr$$

$$= (p-c)\frac{2\pi\rho}{k^2}\int_p^\infty t^{-\varepsilon}(t-p)\,dt$$

$$= \frac{2\pi\rho}{k^2}(p-c)\left[\frac{t^{2-\varepsilon}}{2-\varepsilon}\Big|_p^\infty - \frac{pt^{1-\varepsilon}}{1-\varepsilon}\Big|_p^\infty\right]$$

$$= \frac{2\pi\rho}{k^2}(p-c)p^{2-\varepsilon}\cdot\left[\frac{1}{\varepsilon-2}-\frac{1}{\varepsilon-1}\right]$$

$$= \frac{2\pi\rho}{(\varepsilon-1)(\varepsilon-2)}\cdot\frac{p-c}{k^2}\cdot p^{2-\varepsilon} \tag{31}$$

a decreasing function of costs c and k.

Profit maximization yields

$$p_M = \frac{\varepsilon-2}{\varepsilon-3}c > c \tag{32}$$

Maximal profits are then, using (31) in (30)

$$\hat{G}_M = \frac{2\pi\rho}{k^2}\frac{(\varepsilon-3)^{\varepsilon-3}}{(\varepsilon-2)^{\varepsilon-1}(\varepsilon-1)}c^{3-\varepsilon} \tag{33}$$

The profit ratio is

$$\frac{\hat{G}_U}{\hat{G}_M} = \frac{9}{2}\varepsilon^{-3}(\varepsilon-1)(\varepsilon-2)^{\varepsilon-1} = \frac{9}{2}\cdot\frac{\varepsilon-1}{\varepsilon}\cdot\left(\frac{\varepsilon-2}{\varepsilon}\right)^{\varepsilon-1}$$

For $\varepsilon = 3$ the right hand expression equals $\frac{1}{3}$ and it decreases with $\varepsilon$. Thus

$$\frac{\hat{G}_U}{\hat{G}_M} \le \frac{1}{3}. \tag{34}$$

## References


Beckmann MJ (1976) Spatial Price Policies Revisited. The Bell Journal of Economics 7, 2:619–630

Beckmann MJ (1989) A Note on Spatial Price Policy when Demand is Linear but Non-uniform. Regional Science and Urban Economics 19:103–106

Beckmann MJ, Ingene C (1976) The Profit Equivalence of Uniform and Mill Pricing. Regional Science and Urban Economics

Fetter F (1931) The Masquerade of Monopoly. Harcourt Brace and Co., New York

Greenhut ML, Ohta H (1972) Monopoly Output Under Alternative Spatial Pricing Techniques. American Economic Review 62, 4:705–713

Greenhut ML, Ohta H (1975) Theory of Spatial Pricing and Market Areas. Duke University Press, Durham, NC

Greenhut ML (1981) Spatial Pricing in the USA, West Germany and Japan. Economica 48:79–86

Hoover EM (1937) Spatial Price Discrimination. Review of Economic Studies 4:182–191

Phlips L (1983) The Economics of Price Discrimination. Cambridge University Press, Cambridge

Singer HWA (1937) Note on Spatial Price Discrimination. Review of Economic Studies 5:75–77

Stevens B, Rydell CP(1966) Spatial Demand Theory and Monopoly Price Policy. Papers of Regional Science Association 17:195–204

CHAPTER 4

# Spatial Markets

In this chapter we consider competitive markets. Monopoly was discussed in chapter 3. Monopolistic competition and oligopoly, perhaps the most interesting case, is treated in chapter 7. When buyers and sellers are at possibly different locations, spatial dimension enters into markets. Miksch (1951) has recognized the following distinctions.

| | | SELLERS | |
|---|---|---|---|
| | | *At points* | *In an area* |
| BUYERS | *At points* | Point markets: Trading centers | Supply areas |
| | *In an area* | Market areas | Continously extended markets |

**Table 4.1.** Classification of spatial markets

The four emerging configurations are denoted as market and supply areas, trading centers and continuously extended markets. Spatial markets are organized to move commodities from points of supply to points of demand. Prices must be such as to make commodity movements profitable. When along a path of commodity flow transactions may take place everywhere, then the price differences between buying and selling locations must (just) cover transportation cost: Koopmans' Principle.

When uniform prices are announced by a firm in its market area, this price must cover marginal cost of production plus transportation to all points in the market area. In fact the market radius R is then determined as that distance at which the uniform price just equals marginal cost of production c plus transportation cost kR.

Under mill pricing with transportation supplied by the seller, prices paid by buyers will rise along paths at the rate of transportation cost. A three dimensional graph depicting this is known as a Launhardt cone (Launhardt's Trichter) (cf chapter 2). In spatial markets with prices well defined everywhere, there can be no cross hauling in any homogenous commodity. For price differences will sustain movement only in one direction.

## Market and Supply Areas

The classical case is that of a market area with supply by one or several sellers located in a single location, while buyers are distributed continuously through an area.

Competition requires the presence of several sellers in the supply location, known as localization. Already in medieval cities the members of each trade or craft chose to or were made to locate together in one street or square. Localization is studied in chapter 8.

The Hakimi-Gülicher theorem, discussed in chapter 5, gives a theoretical reason for suppliers to be attracted to nodes in a transportation network. In a central place system, traders are localized in cities of various rank or size (chapter 12).

A natural questions concerns the size of a market area and the number of its suppliers in competitive market equilibrium. For this one must specify demand and the cost structure of firms. Throughout we will make the traditional assumptions of location theory that production cost consists of a fixed cost F and a constant (average or marginal) unit cost c. Here we will add the assumption of a limited capacity m for a firm's output z

$$0 \le z \le m .$$

The demand function is typically assumed to have the same functional form f(p) at all locations, although the density of customers ρ(r) may vary with location r. The first question would be how many firms are needed to supply a market area of given size (radius R) and how the market price at the supplier location is determined. For firms of different size and unit cost, this question is studied in the context of localization (chapter 8). Here we consider the simpler scenario of firms with identical unit cost c , fixed cost F and capacities m.

For recovery of fixed cost the market price must be above unit cost c, and this is possible only when all firms in the market operate at full capacity, so that the demand curve intersects the supply curve of each firm and of the industry in its vertical segment (figure 4.1 p. 49).

When profits have been eliminated through free entry, price exceeds unit cost c by just the fixed cost F allocated to sales m. The number of firms must be such that demand and aggregate supply intersect at the average cost level $c+\frac{F}{m}$ (figure 4.1). The number of firms in competitive equilibrium increases with market size and customer density and decreases with the average cost $c+\frac{F}{m}$.

Consider now supply areas. For agricultural production, land use intensity and price to the buyer in the central market are studied in chapter 11 on land use. In a supply area in which output is produced at a constant rate s per area possibly depending on the market price but independent of transportation costs, the relationship between market area and market prices obtained, simply by equating aggregate supply and demand.

## Competition with Uniform Pricing

Market area analysis assumes special features when suppliers are numerous and close enough for their market areas to overlap and competition to exist everywhere. This requires uniform pricing.

The firms competing with a given one in its location are all within the radius of a market area

$$\frac{p-c}{k}=R$$

which at competitive price p and identical costs c and k is the same for all suppliers.

Since they share the market equally (in the absence of special consumer preferences) (gross) profits from a market area of radius R are $\frac{1}{n}$ of what they are for an exclusive market are of the same size. And this must cover fixed costs F. The larger the number of competing firms or their density (number per area) the higher must be the price needed to earn these fixed costs.

Is there a unique equilibrium price, market radius and density of firms?

As it turns out for each market price, taken as given by the industry, there exists an equilibrium density of firms and the lower the price, the lower the density, i.e. the fewer the number of firms. Looked at the other way the more firms enter the spatial marker, the higher they push the price. In fact, the maximum density of firms is realized when the price reaches its monopoly level, a rather paradoxical conclusion (mathematical note 4.2).

## Trade Centers

When demand and supply are both located at points or trade centers, competitive equilibrium can be described succinctly by three conditions.

(1) Local excess demand, i.e. local demand minus local supply depends only on local price

$$q_i = q_i(p_i)$$

(2) Excess demand is covered by net imports.

(3) Price differences between trade centers do not exceed transportation costs and are equal to them wherever the commodity is traded. Thus price at receiving location equals price at exporting location plus transportation cost and price at any location is less than or equal to price at any other location plus transportation cost. Condition (3) was first made explicit by Koopmans in his famous transportation model (Koopmans 1949). Competitive equilibrium among trade centers can be elegantly described as the maximization of the "social surplus" equal to the consumers' surplus plus producers profits (Samuelson 1952). (On consumers' surplus cf mathematical note 15.1).

If the spatial market is fully connected, i.e. any two points are linked by flows in whatever direction, then any exogenous change in excess demand in one location is transmitted to all locations: the same price increase affects all. But in equilibrium it is also possible that subsystems emerge with flow linkages within but not between subsystems. As long as price changes in one subsystem do not exceed the limit set by condition (3) the other subsystems remain unaffected (Samuelson 1952).

## Spatial Equilibrium in Labor Markets

Besides agricultural supply areas, labor supply areas or commuter markets are of particular interest. Farmers live on their land or in villages adjacent to their acreage. All over the world small merchants reside in or above their working premises, both for security and for convenience. In the early days of industrialization workers were settled next to the factories, particularly when steady working habits had yet to be developed. It is modern urbanized mankind that has escaped from this trap of living where one works. Cheap transportation and above all the mobility of the automobile have enabled modern man to choose where to live from a typically wide array of residential areas. Even when housing markets are tight or non-functioning, say because of rent control, one is not compelled to work right next door. Commuting between residence and workplace is thus a normal pattern of modern life.

Labor is not homogeneous, nor are potential employers. While in a wider sense commuters to a particular firm come from a supply area there is overlap between supply areas of different employers, even for the same type of labor. In fact, even cross hauling is not uncommon. Thus, spatial equilibrium in labor markets has special features to be examined now.

Decision making by workers and employers will depend on the institutional arrangements in spatial labor markets. Analogous to mill and uniform pricing in goods markets, workers will receive either a flat wage, regardless of commuting distance, or a basic wage plus transportation allowances (or plus free transportation, arranged by the company – as is customary in some Third World countries).

When wages are uniform, so that commuting cost is borne by the employee, an incentive exists for the employee to seek the nearest employer for his/her type of labor. With transportation allowances the employer seeks out workers from the nearest residential areas. In both cases supply areas emerge whose radius is as small as possible. When net wages are at or near subsistence level, commuting cost must in effect be paid by the employer. When labor supplies are tight and wages high, competitive pressure on firms will force them to offer flat wages, so that the worker pays the commuting cost.

Overlap is minimal, and thus labor supply areas are as small as possible, when working conditions are similar so that employees are (almost) indifferent among employers. When worker quality is sufficiently uniform so that employers are (almost) indifferent among employees the same happens. The more heterogeneity at either end, the greater the overlap and the larger the (overlapping) labor supply areas.

## Continuous Space Markets

An interesting case is that where a commodity is both produced and consumed over continuously extended areas, as for coal, wheat, and certain other agricultural products. The movements of the commodity from supplier to demander may then be described by a flow field and the spatial prices by sets of curves linking points of equal price, so called iso-tims. The conditions of equilibrium are easily described and are the same as given for point markets. Local excess demand density depends only on local price. Local excess demand, appearing as a sink density, is satisfied by net imports, and excess supply, appearing as a source density, is disposed of by exports. Commodity flows are in the direction of the price gradient, and the price gradient (in absolute value) can never exceed transportation costs and is equal to transportation costs wherever a flow occurs.

These intuitive conditions determine spatial equilibrium. The fact that flows must follow a gradient field implies that the only possible flow patterns are of three types: laminar flow with no singularities, flows emerging from a center, and flows converging to centers. The emerging case represents market areas, the converging case supply areas. Both may occur in different parts.

When iso-tims are sufficiently far apart, the flow system is interrupted and spatially separated submarkets emerge. (For details see mathematical note 4.4).

## Mathematical Note 4.1
## Number of Firms and Price in a Common Market Area

In chapter 2 it was tacitly assumed that a market area has at its center a single supplier. This need not be so, e.g. not in the case of a localized industry or in the setting of a central place system. It is of interest to study competition among several firms located together and serving the same market area. Why several firms?

The most obvious reason for that would be limited capacities (due to limited physical facilities or managements' limited span of control). Let m be this capacity, assumed to be equal to all firms for simplicity, and let there be n firms of this equal size.

Equating demand from the market area and supply nx of the n firms, $x \le m$

$$nx = \int_0^R f(p+kr)\rho(r)dr$$

A competitive price sufficient to cover each firm's fixed cost F exist only if all firms produce at capacity.

**Fig. 4.1.** A representative firm

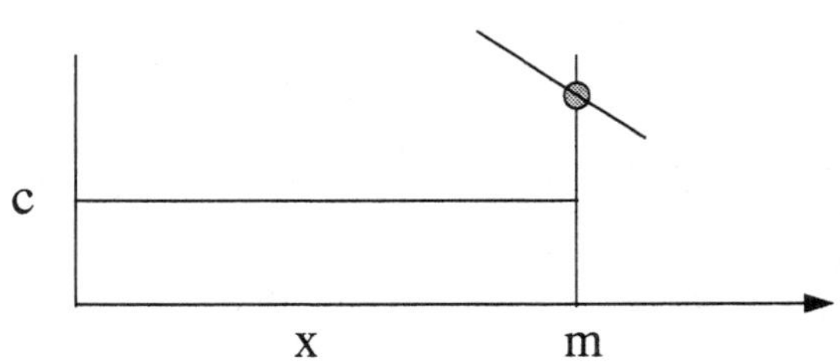


so that

$$(p-c)m = F \tag{1}$$

For linear demand

$$a - p \tag{2}$$

competitive market equilibrium in this market means

$$n \cdot m = \int_0^R [a - p - kr] 2\pi\rho(r) dr \tag{3}$$

with

$$R = \frac{a - p}{k} \tag{4}$$

and

$$p = c + \frac{F}{m} \tag{5}$$

The number of firms is then determined as

$$n = \frac{1}{m}[(a - p) - k\bar{r}]M \tag{5}$$

where $\bar{r}$ is average distance, and M total population in the market area. For constant population density

$$\bar{r} = \frac{2}{3}R \tag{6}$$

in a circular market and

$$M = \pi\rho R^2 \tag{7}$$

then

$$n = \frac{1}{m}\left[a - c - \frac{F}{m} - \frac{2}{3}kR\right]\pi\rho R^2$$

and

$$R = \frac{a - c - \frac{F}{m}}{k}$$

determined by (4).

Thus finally

$$n = \frac{1}{m}\frac{\pi\rho}{3k^2}\left[a - c - \frac{F}{m}\right]^3 \tag{8}$$

This number increases with population density ρ (and the demand intercept a and decreases with capacity and all costs, as one should expect.

### Mathematical Note 4.2
### Competition in Overlapping Market Areas, Uniform Pricing

Market areas can overlap under uniform pricing. Assume that buyers are indifferent among suppliers as long as their prices are the same. At a given competitive price $p$, a supplier chooses its market radius to be the maximal one for covering marginal cost

$$R = \frac{p - c}{k}$$

Let $\lambda$ be the density of suppliers, i.e. the number of suppliers per area. Those that can offer their product to a given customer are all located within the same distance of $R = \frac{p-c}{k}$ and their number is therefore

$$n = \pi\lambda R^2 = \pi\lambda\left(\frac{p-c}{k}\right)^2 \qquad (1)$$

A representative sellers' gross profits are therefore $\frac{1}{n}$ of those he obtains as a single seller

$$\frac{f(p)\int_0^R [p - c - kr]\rho(r)dr}{\pi\lambda R^2}$$

and for uniform density

$$\rho(r)dr = 2\pi\rho dr \qquad (2)$$

$$= f(p)\frac{\rho}{\lambda}\frac{k}{3}\frac{R^3}{R^2} = \frac{f(p)}{3}\cdot(p-c)\frac{\rho}{\lambda} \qquad (3)$$

Net profits are then

$$G = \frac{1}{3}f(p)(p-c)\frac{\rho}{\lambda} - F \qquad (4)$$

If these profits are negative, firms will drop out, if positive, new ones will enter until ($\lambda$ is large enough wipe out profits. Under strict price taking there will be no price cutting (or raising). So a given initial price will stay and the density (of firms will adjust to it. From (4), setting $G = 0$

$$\lambda = \frac{\rho f(p)(p-c)}{3F}$$

There is thus no unique equilibrium density and price, but a whole range of prices for densities (within the range

$$\lambda_{min} = \frac{1}{\pi R^2} \le \lambda \le \lambda_{max} \qquad (5)$$

Here $\lambda_{max}$ is determined by

$$\lambda = \frac{\rho}{3F} \max_{p} f(p) \cdot (p - c) \tag{6}$$

and the right hand side is maximized for the non-spatial monopoly price, maximizing profit margin p – c times sales f(p).

For rectangular demand

$$f(p) = \begin{cases} 1 & p \leq a \\ 0 & p > a \end{cases}$$

the relationship between density and price is linear

$$\lambda = \frac{\rho}{3F}(p - c)$$

$$p = c + 3F\frac{\lambda}{\rho} \tag{7}$$

and for linear demand

$$f(p) = a - p$$

it is quadratic

$$\lambda = \frac{\rho}{F}(a - p)(p - c) \tag{8}$$

with

$$\lambda_{max} = \frac{\rho}{F}\frac{(a - c)^2}{4} \tag{9}$$

at a monopoly price

$$p = \frac{a - c}{2} \tag{10}$$

Suppose now that prices are not rigidly constant but firms respond to profits by attempting to increase market shares through slightly undercutting their competitors' prices. This is, strictly speaking, Bertrand-Edgeworth oligopoly. It brings prices down to the level, where at given density λ profits are wiped out

$$G = \frac{f(p)}{3}(p - c)\frac{\rho}{\lambda} - F = 0 \tag{4a}$$

Here it is price adjustment that is the faster process than that of firm's entry or exist – a reasonable hypothesis. Once more a unique equilibrium price is not determined but one for each density (within the range (5)).

Suppose that both happens: In response to profits, prices are cut and firms enter. The momentum of entry may increase numbers beyond those supported by current prices. In the rectangular demand case, prices must then be raised again to accommodate the higher density. For linear (or flexible demand in general) prices will rise, if they are below the monopoly price level

$$\arg\max_{p} f(p)(p-c) \tag{6a}$$

and fall if they are above. In the first case and for rectangular demand, the paradoxical effect is that an increase in the density of firms leads to higher prices and in any case a maximum density of firms can exist only at monopoly prices.

The minimum density is that at which the market areas no longer overlap. It will be explored in chapter 7, as Bertrand Oligopoly. The Bertrand price is consistent with (4a)

$$G = 0 \text{ and } \lambda = \frac{1}{\pi R^2}$$

## Mathematical Note 4.3
## Equilibrium of Trade Centers

How are prices, net exports and commodity flows between points – trade centers – determined when excess demand functions are given for each of the trade centers and transportation costs are known between any pair of them? Following Samuelson (1952) we postulate that in a system of competitive markets the social surplus, the sum of consumers and producers surplus (profits) is maximized. (For the consumer surplus cf. note 15.2)

Let $q_i\,(p)$ be the excess demand function in location i and consider

$$\int_0^{q(p)} p(t)dt = u(p) \qquad \text{(say)} \tag{1}$$

where p(q) is the inverse function to q(p). The social surplus for the system of trade centers is then

$$\sum_i u(p_i) - \sum_{ij} k_{ij} x_{ij} \tag{2}$$

to be maximized subject to the constraints on commodity flows $x_{ij} \geq 0$ from i to j

$$\sum_j \left(x_{ji} - x_{ij}\right) = q_i\left(p_i\right) \tag{3}$$

The left hand sum in (3) is inflow from other locations j minus outflow to other locations j, hence imports and these must fill the excess demand $q_i$. In view of (1), (2), (3) the problem becomes an unconstrained maximum problem of a concave function in non-negative variables $x_{ij}$

$$\underset{x_{ij} \geq 0}{\text{Max}} \sum_i \int^{\sum_j (x_{ji}-x_{ij})} p_i(t)dt - \sum_{ij} k_{ij} x_{ij} \tag{4}$$

The maximand is (strictly) concave when the excess demand functions $p_i(q)$ are (strictly) decreasing functions. (4) is solved by

$$x_{ij}\begin{Bmatrix} = \\ \geq \end{Bmatrix} 0 \iff p_j - p_i \begin{Bmatrix} < \\ = \end{Bmatrix} k_{ij} \quad (5)$$

Conditions (3) and (5) together determine the commodity flows $x_{ij}$ and prices $p_i$ of the system. They are unique when the maximand is strictly concave, that is all excess demand functions are strictly decreasing.

When all trade centers are connected by flow vectors in whatever direction, then any price change in one location, caused e.g. by a local shift in its excess demand curve is transmitted to all other locations.

Of course, the equilibrium flow system need not be connected but may break up into several isolated sub systems.

An interesting example of pricing in trade centers is the fact observed by Lösch (1940) that interest rates rise with distance between lender and borrower, and thus tend to be higher the farther removed the location is from the great financial centers, predominantly in New York.

This is not due to any transportation costs for funds but to the lenders perception of increased risk at increased distances.

### Mathematical Note 4.4
### Spatial Equilibrium in Labor Markets

We consider first an efficient spatial allocation of labor. The objective to be maximized is the total product of labor minus commuting cost. The production technologies considered are of the fixed coefficient type: firm k (the firm in location k) may employ labor of a given type at constant marginal productivity $w_k$ up to an (employment) capacity limit $c_k$. Our activity variable is $x_{ik}$, the number of commuters (of this type of labor) from residences i to firms k. Let $r_{ik}$ be the commuting cost, a measure of distance, from i to k.

$$\max_{x_{ik} \geq 0} \sum_k v_k \sum_i x_{ik} - \sum_{ik} r_{ik} x_{ik} \quad (1)$$

The constraints are as follows: availability of workers in residential location i

$$\sum_k x_{ik} \leq a_i \quad (2)$$

employment capacity of firm k

$$\sum_i x_{ik} \leq c_k. \quad (3)$$

We now study the Koopmans efficiency conditions. Consider the Lagrange function

$$L = \sum_{ik} (v_k - r_{ik}) x_{ik} + \sum_i \lambda_i \left( a_i - \sum_k x_{ik} \right)$$
$$+ \sum_k \mu_k \left( c_k - \sum_i x_{ik} \right)$$

It involves dual variables or efficiency prices

$$\lambda_i = 1, \ldots m \text{ and}$$
$$\mu_k = 1, \ldots n$$

associated with residential and firm locations respectively. The Koopmans efficiency conditions are obtained by maximizing L with respective to $x_{ik} \geq 0$ and minimizing with respective to $\lambda_i \geq 0$, $\mu_k \geq 0$.

They are

$$x_{ik} \begin{Bmatrix} = \\ \geq \end{Bmatrix} 0 \Leftrightarrow \qquad v_k - r_{ik} \begin{Bmatrix} < \\ = \end{Bmatrix} \lambda_i + \mu_k \tag{4}$$

where $\lambda_i$ is a premium paid to workers residing in i and $\mu_k$ a quasi-rent earned, per employee, by a firm in location k.

Now,

$$\lambda_i \begin{Bmatrix} = \\ \geq \end{Bmatrix} 0 \Leftrightarrow \sum_k x_{ik} \begin{Bmatrix} < \\ = \end{Bmatrix} a_i \tag{5}$$

$$\mu_i \begin{Bmatrix} = \\ \geq \end{Bmatrix} 0 \Leftrightarrow \sum_i x_{ik} \begin{Bmatrix} < \\ = \end{Bmatrix} c_i \tag{6}$$

The right part of (4) may also be written

$$\lambda_i = \max_k \left( v_k - r_{ik} - \mu_k \right) \tag{4a}$$

or

$$\mu_k = \max_1 \left( w_k - r_{ik} - \lambda_i \right) \tag{4b}$$

The two equations (4a), (4b) represent two possible institutional arrangements for implementing competition in spatial labor markets. In (4a) the market sets wage rates $v_k$-$\mu_k$ in the various work places. Workers then choose employers by comparing net wages after commuting costs and taking the best offer

$$\lambda_i = \max_k \left( w_k - r_{ik} \right) \tag{4c}$$

where

$$w_k = v_k - \mu_k$$

is the market wage. This is in effect a sellers' market, most likely to emerge in the event of a labor shortage

$$\sum_k c_k > \sum_i a_i$$

Alternatively, competition may fix workers net earning to be $\lambda_i$ above subsistence levels and employers then seek out labor at lowest marginal cost.

$$v_k - \mu_k = \min_i \left(\lambda_i + r_{ik}\right) \tag{4d}$$

and thus maximize rent

$$\mu_k = \max_i \left(v_k - \lambda_i - r_{ik}\right)$$

$$= v_k - \min_i \left(\lambda_i + r_{ik}\right) \tag{4b}$$

This is the appropriate scenario for a buyers' market with unavoidable unemployment when

$$\sum_k c_k < \sum a_i \; .$$

Interestingly enough in developing countries like India where buyers markets are common, employers often provide transportation, and thus incur commuting costs $r_{ik}$ in addition to the subsistence or legal minimum wages $\lambda_1$.

When unemployment is widespread, i.e., occurring in all residential locations

$$\sum_k x_{ik} < a_i$$

then

$$\lambda_i \equiv 0$$

which must be interpreted as net wages $\lambda_1$ being at the subsistence or legal minimum level.

With $\lambda_1$ uniform the employers objective becomes that of hiring so as to minimize total commuting cost, which means hiring only from the nearest residential location.

In any event competition determines both $\lambda_1$ and $\mu_k$ and they are consistent provided (4) holds implying both (4a) and (4b). The set of firm locations k in which labor's net wage is maximized = $\lambda_i$ is the labor market area of residential location i. Conversely, the set of residential locations i where labor will maximize employers rent $\mu_k$ is the labor supply area of firm k. Observe, however, that when aggregate labor demand and supply are matched

$$\sum_i a_i = \sum_k c_k \tag{7}$$

the levels of employers rent $\mu_k$ and labor net wage $\lambda_1$ are no longer uniquely determined since a (positive or negative) constant c may be added to all $\lambda_i$ provided it is subtracted from all $\mu_k$. (This leaves a range of indeterminacy for c limited by the requirement that all $\lambda_1 \geq 0, \mu_k \geq 0$.

$$0 \leq \min_k \mu_k - c$$

or

$$-c \leq \min \mu_k$$

Notice also that in the balanced case just considered the $v_k$ are irrelevant since at full employment the allocation problem is equivalent to the Koopmans transportation problem.

$$\min_{x_{ik} \geq 0} \sum_{ik} r_{ik} x_{ik} \tag{8}$$

such that

$$\sum_k x_{ik} = a_i$$
$$\sum_i x_{ik} = c_k$$

**Mathematical Note 4.5**
**Spatial Equilibrium in a Continuous Market**

The natural way to model a market with continuously distributed supply and demand is to let the flow of the commodity traded be visualized as a continuos flow field $\phi(x_1, x_2)$ or $\phi(x)$ (Beckmann 1952, Beckmann and Puu 1985). The local excess supply which is exported is then the source density or "divergence" of the flow field

$$\operatorname{div} \phi = \frac{\partial \phi_1}{\partial x_1} + \frac{\partial \phi_2}{\partial x_2} \tag{1}$$

Let q (x) be the density of excess demand in location x, the negative of excess supply. Then

$$\operatorname{div} \phi + q = 0 \tag{2}$$

is the "divergence law" stating that excess supply constitutes the source density of the flow field.

To transport the commodity at minimum cost as required by market efficiency, prices change locally along shortest paths by just enough to recover transportation cost. This means that flow moves in the gradient direction of a potential function, the potential being price $\lambda$

$$\phi \sim \operatorname{grad} \lambda \tag{3}$$

and that

$$\begin{aligned} |\operatorname{grad} \lambda| &= k \quad \text{for } \phi \neq o \\ |\operatorname{grad} \lambda| &\leq k \quad \text{for } \phi = o \end{aligned} \tag{4}$$

Combining (3) and (4)

$$\operatorname{grad} \lambda = k \frac{\phi}{|\phi|} \tag{5}$$

This is the so-called *gradient law.*
(2) and (5) determine the flow field $\phi$ and its potential $\lambda$.

At the regional boundaries the net outflow from the region must be specified. In a self-sufficient region the total net outflow must be zero.

$$\oint \phi_n ds = 0 \tag{6}$$

A stronger condition is that the net outflow should vanish everywhere on the boundary

$$\phi_n = 0 \tag{7}$$

When excess demand depends on price

$$q = q(x, p) \tag{8}$$

then the potential $\lambda$ is the price

$$\lambda(x) = p(x) \tag{9}$$

**Mathematical Note 4.6**
**Existence of Spatial Markets**

Under what conditions can a spatial market exist, when can demand and costs sustain a firm selling to customers at a distance?

We examine first the more elementary question when is a spatial division of labor feasible? Let $c_0$ be unit cost of in-house production, q quantity consumed in a household. Comparing costs in a spatial market with in-house cost

$$\begin{aligned} &\min_R \left( c + k\overline{R} + \frac{F}{\pi\rho q R^2} \right) \le c_0 \\ &c_0 - c \ge \min_R \left( \frac{2}{3} kR + \frac{F}{\pi\rho q R^2} \right) \end{aligned} \tag{1}$$

The market radius for minimal unit cost is determined by

$$0 = \frac{2}{3}k - \frac{2F}{\pi\rho q R^3} \tag{2}$$

or

$$R^3 = \frac{3F}{\pi\rho q k} \qquad R = \left( \frac{3F}{\pi\rho q k} \right)^{\frac{1}{3}} \tag{3}$$

achieving unit costs

$$k\frac{2}{3}\sqrt[3]{\frac{3F}{\pi\rho q k}} + \frac{F}{\pi\rho q}\left( \frac{3F}{\pi\rho q k} \right)^{-\frac{2}{3}} = \left( \frac{3F}{\pi\rho q} \right)^{\frac{1}{3}} k^{\frac{2}{3}} \tag{4}$$

so that the condition (1) becomes

$$\left(\frac{3F}{\pi\rho q}\right)^{\frac{1}{2}} k^{\frac{2}{3}} \le c_0 - c$$

$$\frac{3F}{\pi\rho q} \cdot k^2 \le (c_0 - c)^3 \tag{5}$$

The left hand is an increasing function of costs and a decreasing function of quantity demanded per household and of household density, as one would expect.

A more sophisticated argument with demand depending on price can be derived from the profit functions of chapter 3. A market can exist, i.e. production by one firm in one location for a market of radius R is profitable at a mill price $p_M$ provided

$$(p - c)\int_0^R f(p + kr)\rho(r)dr \ge F$$

For linear demand and constant population density a firm is profitable using mill or uniform pricing whenever

$$\frac{(a - c)^4}{Fk^2} \cdot \rho \ge \frac{256}{9\pi} \doteq 9.054 \tag{6}$$

using (3.4.9).

Generally for some pricing system p(r) and market radius R

$$\int_0^R [p(r) - c - k(r)]f(p(r))\rho(r)dr \ge F$$

with market radius determined by

$$c + kR = p(R) = a$$

linear demand

$$\int_0^R [\pi - c - k(r)][a - p(r)]\rho(r)dr \ge F,$$

and perfect discrimination

$$p(r) = \frac{1}{2}(a + c + kr),$$

and constant population density

$$\rho(r) = 2\pi\rho r$$

one obtains the less restrictive condition

$$\frac{(a - c)^4}{Fk^2}\rho \ge \frac{24}{\pi} \doteq 7.639 \tag{7}$$

Recall that for linear demand functions ρ has the dimension $\frac{1}{\text{price}}$ so that the expressions in (6), (7) are dimensionless.

A rectangular demand function with reservations price a lies above the linear demand function with intercept a so that the survival condition under uniform and perfectly discriminating pricing (which are identical here) is

$$\frac{(a-c)^3}{Fk^2}\rho \geq \frac{3}{\pi} \doteq 0.955 \tag{8}$$

Recall that for rectangular demand ρ is dimensionless.

Notice that the cost criterion (5) for the existence of a spatial market becomes identical to that for rectangular demand (8) if the reservations price a is set at the level $c_0$ of cost for in-house production.

To meet the criterion for survival of a firm, that is, the existence of a market the spread between demand intercept a and marginal costs c and the density of consumers ρ must be large relative to transportation cost k and fixed cost F by a factor depending on market organization. But the sensitivity to transportation cost is stronger than that to fixed cost and is largest with respect to the spread a–c.

With uniform pricing, constant population density but arbitrary demand functions, the condition for survival of a firm is

$$\underset{p}{\text{Max}}\, f(p) \int_0^{\frac{p-c}{k}} [p-c-kr]\, 2\pi\rho\, dr \geq F$$

or

$$\rho f(p) \frac{(p-c)^3}{k^2 F} \geq \frac{3}{\pi} = .955 \tag{9}$$

for some p.

## References


Beckmann MJ (1952) A Continuous Model of Transportation. Econometrica 20:643–660

Beckmann MJ, Puu T (1985) Spatial Economics: Density, Potential, and Flow. Elsevier, Amsterdam

Fox K, Sengupta JK (1969) Economic Analysis and Operations Research. North-Holland, Amsterdam

Gosh A, Chakroborti A (1973) Programming and Interregional Input-Output Analysis: An Application to the Problem of Industrial Location in India. Cambridge University Press, Cambridge

Heady EO (1958) Linear Programming Methods. Iowa State College Press, Ames

Henderson JM (1958) The Efficiency of the Coal Industry; An Application of Linear Programming. Harvard University Press, Cambridge

Lefeber L (1958) Allocation in Space, Production, Transport and Industrial Location. North-Holland, Amsterdam

Miksch L (1951) Theorie des räumlichen Gleichgewichts Weltwirts. Archiv 66:55–47

Samuelson PA (1952) Spatial Price Equilibrium and Linear Programming. American Economic Review, June, p 283–303

Takayama T, Judge G (1964) Spatial Equilibrium and Quadratic Programming. Journal of Farm Economics 46:1

# CHAPTER 5

# Location of an Economic Activity

The location of firms is at the very heart of location theory. Rather than firms, one should consider plants however, for a firm may have several plants (chapter 6), and plants are but one type of production facility. So it is common now to speak of the location of an economic activity. This includes services and distribution, but not residential housing or agriculture, two types of land intensive economic activities that are usually considered in the context of land use (chapter 11).

When entrepreneurs are asked about their locational decisions, in questionnaires say, they tend to list a number of factors considered important such as

- Space for (later) expansion
- Available labor, with needed skills and at low cost
- Low taxes or even subsidies
- Infrastructure, in particular
- Accessibility through transportation
- Amenities

These are all important in their contribution to (present and future) profitability, which after all is an absolute must for long term survival in a market economy.

Economic theory suggest in fact that the best chance of survival is gained by maximizing profits. This goal may sometimes be spurned, when entrepreneurs exercise personal preferences, as long as profitability is maintained. Thus Henry Ford chose Detroit over Chicago when locating his first automobile assembly plant, and US Steel selected Gary, Indiana, over nearby Chicago for political reasons. Japanese business firms operating in Germany have an unexplainable attraction to Düsseldorf rather than Frankfurt and there is a well-known preference for San Francisco over Los Angeles, or for Geneva over Zurich, shown by various business firms.

Following economic tradition we will assume as a working hypothesis that locational choice is guided after all by the objective of maximum profits.

## Orientation

Now profits equal revenue minus costs, and we must next distinguish between situations where either revenues or costs are more sensitive to location. Costs in turn are the sum of production and transportation costs. Classical location theory starts out with the case where both revenues and production costs are constant and only transportation costs depend on location, so that profit maximization reduces to the minimization of transportation costs, also known as transport orientation.

The simplest and perhaps even the most important case is that where a single material input is converted into one output, and the proportions are fixed. The relevant question with regard to transportation cost is then whether it costs more to move the input or the output.

This is often, but not exclusively, a question of weight. A reduction of weight (or when weights are unchanged, of bulk) results in cheaper transportation of the product compared to the raw material, causing the processing activity to be located at the resource deposit. This case is known as *resource orientation* (table 5.1)

When the product is heavier – or, without change of weight, bulkier – it is cheaper to transport the raw material to the production site, and we have *market orientation.* Following are some examples of resource orientation.

- Smelting and refining ores: copper, tin, zinc and similar metals
- Hydro-power
- Electric power production from lignite
- Pressed coal from coal dust
- Boards from trees: saw mills
- Paper from wood chips: paper mills
- Bricks from clay: brick yards
- Porcelain
- Canning and preserving fruits and vegetables
- Sugar refining
- Tobacco curing
- Coffee drying (not roasting)
- Cotton combing
- Wool washing and carding
- Hide tanning
- Slaughtering and meat packing (after refrigerated transportation was introduced)
- Bottling of mineral water
- Bottling of quality wines
- Butter and cheese: dairies

Before considering market orientation, another possibility must be examined. When weight or bulk are not changed greatly in the processing, economies of scale may be utilized at points where raw material streams converge and before the product streams diverge. This happens typically at trans-shipment points where the material or product is loaded from one mode of transportation to another, so that in fact the reloading may be combined economically with the processing, sometimes called "milling in transit" (Hoover 1948). Examples of this are

- Milling grain into flour at the transit from ship to rail e.g. in Minneapolis and Buffalo
- Refining crude oil to gasoline at seaports or pipeline terminals

**Table 5.1.** Types of orientation

| Costlier to move | Input | Output |
|---|---|---|
| Orientation | Resource | Market |

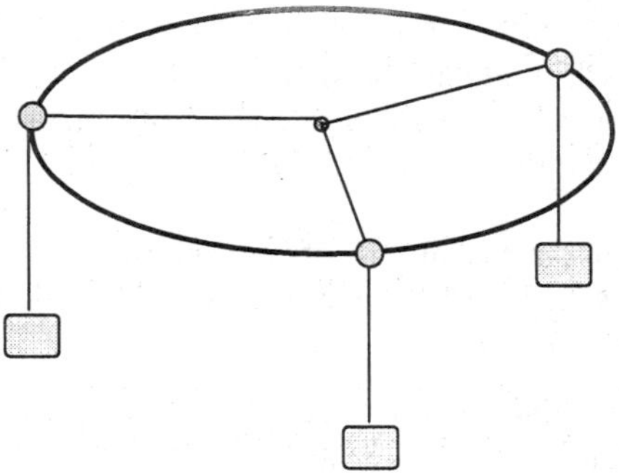

**Fig. 5.1.** Varignon machine

The case of an output being produced from two inputs in fixed proportions has attracted an undue amount of attention in the location theoretical literature. The example one has in mind is that of producing iron for a single market from coal and ore. The transportation cost minimizing location need not be in any of the three given points. If it falls into the interior of the "Weberian triangle" spanned by the two inputs and one market location it is at the point of equilibrium between the three forces pulling towards the three corners with strengths proportional to the weights that must be moved. This condition was first discovered by W Launhardt (1882) who studied this "Weber problem" before Alfred Weber (1905). When the three weights are equal, the "Weber problem" becomes the "Steiner problem", well known to mathematicians. In its general form the object is the linking of any given set of points by a shortest spanning tree with nodes added as needed.

In the three point case, the transport cost minimizing location can be visualized by a device of Varignon in which three strings carrying the respective weights are allowed to pull a knot into an equilibrium position (fig 5.1)

When one force exceeds the other two, or nearly so, then the equilibrium point of the forces will be pulled into a corner of the triangle, i.e. into a resource or the market location. It is also possible that both resource locations become producing locations, e.g. when the transportation system utilizes its carriage capacity for hauling coal in one direction and ore on the return trip, as is done in Russia between the Donez coal basin and a not too distant ore deposit.

When proportions are not fixed, the analysis of plant location must be broadened. Consider a move of the plant by a unit distance away from the market towards a resource deposit. It will not only change transportation cost by the differences in product and input weights but also modify production cost as a result of changes in the local wage rate and the delivered price of the resource input which then give rise to adjustments in the proportions of the inputs used. If wage rates and resource transportation cost both fall in the direction to the deposit, then the result is resource orientation again, provided the finished product weighs less than the input when, at local prices, the cost minimizing proportions are used. When wages and delivered resource price vary in opposite directions, and the wage gradient is strong enough, a location between resource deposit and market is possible as the cost minimizing choice. This should then be called labor orientation.

Unfavorable locations with high transportation costs for inputs or outputs can be competitive only by having lower production costs, which means typically lower labor cost, *labor orientation.* These are the result either of lower wages or of the possession of special skills that have been embodied in immobile labor through local traditions. Such labor intensive activities are often organized as cottage industries employing predominantly women and children at home as in

- Lace making in the coal district of Pennsylvania
- Toys and musical instruments in the mountain valleys of central Germany (Thüringer Wald, Erzgebirge)

A modern type of labor orientation are assembly plants for electronic products in low wage countries, located near airports through which both the material and the product are moved.

## Market Orientation

Turn now to *market orientation,* which according to C Harris (1954) has gained increasing importance with modern technology. As a device for saving transportation cost it can motivate firms directly only when these incur the transportation costs as under uniform pricing. For mill pricing transportation cost affects firm behavior through its effect on sales revenue. Competitive pressure will also motivate firms to seek locations offering lowest price including transportation cost or maximum convenience to buyers. In fact when demand curves are linear, there is a mathematical equivalence in the role of transportation cost under either pricing system (mathematical note 5.2). In general, under mill pricing, market orientation must be interpreted not as a way of minimizing transportation cost but of maximizing revenues. Since product markets are usually dispersed, market orientation, unlike resource orientation, does not yet fix the optimal location but opens the question where in regard to the spatially distributed demand a supplier should locate.

A simple answer is possible for identical linear demand functions: the principle of the *median.* When customer locations extend along a line, the transportation cost minimizing or revenue maximizing location is at the median with equal weights or quantities moving in either direction. Under a Manhattan metric the median is sought for both the North-South and the East-West directions (say). Examples of market oriented industries include most consumer goods such as

- Distribution of food stuffs, drugs and cosmetics
- Soft drink bottling from syrups
- Bottling of bulk wines(ordinary wine)
- Production of pots, pans and other household goods
- Some home furnishings
- Restaurants
- Tailoring
- Assembly of household durables
- Book stores
- Automobile assembly plants
- Services to households
- The performing arts

Firms in the garment industry, while close to markets in a general sense, are not simply market oriented but are concentrated near a few major market places (chapter 9) through the need for rapid communication in the quickly changing world of fashion. A similar role is played by access to information and new knowledge in the making of computers and subsidiary products (the Silicon Valley phenomenon). Both types of economic activity may be described as *knowledge and information oriented.*

Returning to production costs as a locational factor, it is sometimes rent rather than wages that dominates, resulting in *rent minimization* as the object of locational choice. More often rent may be considered as a restraining factor in market orientation. Thus restaurants and book stores seek access to consumers by locating in busy areas. Since they cannot afford the main streets they seek out less frequented side streets, while department stores, and art galleries need the most accessible locations. In seeking market areas subject to an affordable rent we have an example of both revenue and cost depending on location so that profit maximization results in neither cost minimization nor revenue maximization, but sometimes revenue maximization subject to a limit on costs.

Market orientation occurs also when markets are not defined by household locations, but have their own identifiable pattern. Thus, book stores, liquor stores (where allowed), cheap restaurants, and stores for juvenile clothing are attracted to the neighborhood of university campuses. (More in chapter 9: Agglomeration).

As a final possibility neither cost nor revenues may be dependent on location, which means a *footloose activity* . These take two forms. Production for a universal market from so-called *ubiquitous* resources (occurring everywhere, at more or less constant cost) of which

- Brick making

is perhaps the only example. Or the activity may not require any localized resources or sell to localized markets at all. This applies perhaps only to such creative activities as

- Painting
- Writing and
- Basic research

Among the footloose we also find the non-producing population of retired persons.

Footloose artists (painters and writers) fall into two classes: poor and rich. The poor will seek out locations offering low costs of living, the rich locations distinguished by an agreeable climate, scenic beauty, cultural amenities, and if possible, tax benefits. Locations of beauty and amenity are also attractive to scientists in basic research. (These absent minded professors are not totally unaware of their environment). The Bay area near San Francisco, San Diego and Los Angeles and southern California in general, as well as Santa Fe have all attracted research centers or "think tanks" for these reasons.

The leisure class of affluent retired persons, by bidding up rents, has in some cases forced artists out of the artist colonies and professors away from their campus neighborhoods when these were sufficiently attractive. Examples include Evanston, Illinois, the location of Northwestern University and Ann Harbor, Michigan, the location of the University of Michigan, as well as Westwood, near the University of California in Los Angeles.

## Location in a Network

When production sites are restricted to those on an existing transportation network, then even without change in the mode of transportation, nodal points, i.e. junctions, in the network become transport cost minimizing attractors simply as a result of being points of equilibrium between weight pulling forces. For, consider a location on a line segment of the transport system and compare the weight that pull in the two directions. Then any adjustment towards the stronger pull serves to reduce transportation cost until a vertex, a junction point on the network is reached. Now consider a "virtual displacement" along one of the joining lines. Once more the stronger pull indicates direction of transport cost saving. A vertex can be a candidate for a transportation cost minimum only if on all virtual displacements the resulting pull is towards the vertex. There may be many such vertex solutions, each representing a "local minimum" to the Weber problem. A direct comparison among them will then yield the global minimum.

That the search for the Weber point can be restricted to the vertices of a network has become known as the Hakimi Theorum although this was independently discovered by H Gülicher (Hakimi 1964, Gülicher 1965).

Nodes in a network constitute not only the points of lowest transportation costs but also those of highest traffic volume for traffic inflow into any node j exceeds the flow on any inbound line $x_{ij}$ and traffic outflow from j exceeds the flow on any outbound link $x_{jk}$. This is relevant to economic activities catering to travellers such as hotels, restaurants, gas stations and tourist attractions. Traditional hotels are attracted to railroad terminals, modern ones to airports and intersections of major highways or to exits of freeways. Restaurants of the gourmet type can exist in smaller cities provided they are on or close to major routes travelled by affluent automobile tourists.

Theme parks and modern tourist attractions are drawn to locations near transportation hubs. Examples are the Disney parks in Anaheim (near the Los Angeles hub), Orlando (a hub for travellers to Florida resorts), in Tokyo and Paris.

### Mathematical Note 5.1
### Launhardt-Weber Model

Let a, b, c (in vector notation) denote the three given locations and x plant location to be determined. Let weights $w_a$, $w_b$, $w_c$ be moved over distances $|x - a|, |x - b|, [x - c]$, the minimand is then

$$\underset{X}{\mathrm{Min}}\; w_a |a - c| + w_b |b - z| + w_c |c - z| \tag{1}$$

or in terms of components

$$w_a\sqrt{(a_1-x_1)^2+(a_2-x_2)^2}+w_b\sqrt{(b_1-x_1)^2+(b_2-x_2)^2}$$
$$+w_c\sqrt{(c_1-x_1)^2+(c_2-x_2)^2} \tag{2}$$

This is a concave function in $x_1$, $x_2$. Necessary and sufficient for an interior minimum is therefore the vanishing of the derivatives

$$0=-w_a\frac{a_i-x_i}{\sqrt{(a_1-x_1)^2+(a_2-x_2)^2}}-w_b\frac{b_i-x_i}{\sqrt{(b_1-x_1)^2+(b_2-x_2)}}$$
$$-w_c\frac{c_i-x_i}{\sqrt{(c_1-x_1)^2+(c_2-x_2)^2}} \quad i=1,2$$

or in vector notation,

$$0=w_a\frac{a-x}{|a-x|}+w_b\frac{b-x}{|b-x|}+w_c\frac{c-x}{|c-x|} \tag{3}$$

$w_a\frac{a-x}{|a-x|}$ is a vector of length $w_a$ pointing from x to a. At the minimum cost location x these three vectors, representing pulling forces, must be in equilibrium.

When weights are too unequal to permit such an equilibrium, the cost minimum falls into a corner of the triangle spanned by the locations a, b, c. It is sufficient but not necessary for this that one weight exceed the sum of the other two.

### Mathematical Note 5.2
### Principle of the Median

Under uniform pricing at given price $p_U$ in a linearly extended market are $r_0 \le r \le r_1$ with customer density $\rho(r)$ consider the profit maximizing location x

$$\underset{X}{\mathrm{Max}}\, f(p_U)\int_{r_0}^{r_1}\left[p_U - c - k|r-x|\right]\rho(r)dr \tag{1}$$

where k is transportation cost per unit commodity per unit distance

c unit production cost
f(p) demand function

since c is independent of location, provided all customers are served in the given market area $r_0 \le r \le r_1$, total production cost $\int c\rho(r)dr$ may be dropped from the maximand resulting in net revenue as the objective function. On dropping the factor f $(p_U)$ and the constant term $p_U\int_{r_0}^{r_1}\rho\, dr$ the result is minimization of transportation cost

$$\underset{x}{\text{Min}}\ k\int_{r_0}^{x}(x-r)\rho(r)dr + k\int_{x}^{r_1}(r-x)\rho(r)dr \tag{2}$$

yielding

$$\int_{r_0}^{x}\rho(r)dr = \int_{x}^{r_1}\rho(r)dr \tag{3}$$

That is x should be the median of the customer distribution $\rho(r)$. It is equal to the mid point when $\rho(r)$ is constant.

When the boundaries $r_0$, $r_1$ may be chosen one has the additional condition

$$\underset{r_0,r_1}{\text{Max}}\, f(p)\int_{r_0}^{x}[p-c-k(x-r)]\rho(r)dr + f(p)\int_{x}^{r_1}[p-c-k(r-x)]\rho(r)dr$$

yielding

$$p-c-k(x-r_0)=0 \qquad p-c-k(r_1-x)=0 \tag{4}$$

so that deliveries are made to the point where marginal revenue $p_U$ equals marginal cost

$$c + k \cdot \text{distance}.$$

Under mill pricing and constant unit cost c, profit maximization requires instead

$$\underset{x}{\text{Max}}\ (p_M - c)\left[\int_{a}^{x} f(p_M + k\cdot(x-r))\rho(r)dr + \int_{x}^{b} f(p_M + k(r-x))\rho(r)dr\right]$$

which for a given mill price $p_M$ reduced to sales maximization.

$$\text{Max}\int_{a}^{b} f(p_M + k|r-x|)\rho(r)dr \tag{5}$$

Assume f to be linear with unit slope coefficient through a suitable choice of the commodity unit

$$f(p) = a - p \tag{6}$$

Then sales maximization (5) is equivalent to the minimization of transportation cost (2) and yields the principle of the median once more.

For arbitrary demand functions f (p) the principle of the median does no longer hold in its straight form, but only with respect to marginal demand

$$\int_{a}^{x} k\, f'(p_M + k(x-r))\rho dr - \int_{x}^{b} k\, f'(p_M + k(r-x))\rho dr = 0$$

Since this marginal demand depends itself on the local price this is not a principle easily applied to non-linear demand.

### Mathematical Note 5.3
### The Hakimi-Gülicher Theorum

When locations must be chosen on a given network, the *Weber problem* of the fixed weights of inputs and outputs to be moved at minimum cost is considerably simplified. Consider a location x on an edge ij of the network (figure 5.3), compare the weights $w_i$ and $w_j$ to be moved to or from i, j to the production location x. Let $\ell_{ij}$ be the length of ij. Transportation costs on this edge are

$$w_i x + w_j\left(\ell_{ij} - x\right) \tag{1}$$

and transportation cost beyond i and j may be treated as given when considering a location on edge ij. Minimizing (1) with respect to x yields

$$x = \begin{Bmatrix} 0 \\ \ell_{ij} \end{Bmatrix} \Leftrightarrow w_i \begin{Bmatrix} > \\ < \end{Bmatrix} w_j \tag{2}$$

while x is arbitrary when $w_i = w_j$ and hence may be chosen at either i or j. x = 0 in equation (2) means location at i and $x = \ell_{ij}$ location at j.

**Fig. 5.3.** Location on an edge ij of a network

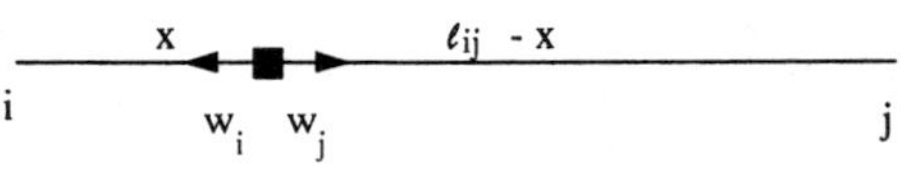


## References

Beckmann MJ (1967) On the Optimal Location of Highway Networks. In: Garrison WL, Marble DF (eds) Quantitative Geography, Part I. Northwestern University Press, Evanston, IL
Bos HC (1965) Spatial Dispersion of Economic Activities. Rotterdam University Press, Rotterdam
Greenhut M (1956) Plant Location in Theory and Practice. University of North Carolina Press, Chapel Hill
Gülicher H (1965) Einige Eigenschaften optimaler Standorte in Verkehrsnetzen. Schriften des Vereins für Sozialpolitik, New Series, Vol 42:110–137
Hakimi SL (1964) Optimum Location of Switching Centers and the Absolute Center and Median of a Graph. Operations Research 12:450–459
Harris CD (1954) The Market as a Factor in the Localization of Industry. Annals of the Anvication of American Geographus 44:315–348
Hoover EM (1948) The Location of Economic Activity. McGraw-Hill, New York
Isard W (1956) Location and the Space Economy. Wiley, New York
Launhardt W (1882) Der zweckmäßigste Standort einer gewerblichen Anlage. Zeitschrift des Vereines deutscher Ingenieure
Moses L (1958) Location and the Theory of Productionb. Quarterly Journal of Economics 72:259–272
Predöhl A (1925) Das Standortproblem in der Wirtschaftstheorie. Weltwirts. Archiv 21:294–331
Sargent FP (1949) Investment, Location and Size of Plant. Cambridge University Press, Campridge
Weber A (1909) On the Location of Industries (translation of Über den Standort der Industrien). University of Chicago Press. Chigago (1929)

# CHAPTER 6

# Multiple Facility Location

Locating multiple facilities of a single firm or agency falls between the location problems of a plant and of an industry.

We distinguish between outlets of a business firm, and the facilities of a public agency. In the latter case the demand for its services may be given and fixed or it may depend on accessibility.

## The McDonald Problem

The best known example of firms with multiple outlets are chains of department stores, of specialty stores and of fast food restaurants. Following Lovell (1970) we shall speak of the McDonald location problem.

The location objective will be profit maximization for the entire chain, but in the short run it can also be a preemption of market areas to forestall rival chains from entering (see below). Both objectives are achieved when the profit maximizing locations are close enough.

How far should facilities be spaced? What is the profit maximizing objective: sales per store, sales per area, or sales per target household? When household density is uniform, the last two objectives are equivalent.

Now, total profits in a region are profits per store times stores in the region, and

$$\text{number of stores in the region} = \frac{\text{regional area}}{\text{market area of store}}$$

Thus

$$\begin{aligned}
\text{total profits} &= \frac{\text{profits}}{\text{store}} \cdot \text{number of stores} \\
&= \frac{\text{profits}}{\text{store}} \cdot \frac{\text{regional area}}{\text{market area of store}} \\
&= \frac{\text{regional area}}{\text{store}} \cdot \frac{\text{profits}}{\text{market area}} \\
&\sim \frac{\text{profit}}{\text{market area}}
\end{aligned}$$

The object should thus be to maximize profits per area or, at constant customer density, profits per customer. For urban markets it seems appropriate to assume Manhattan geometry in calculating distances between stores and customers, rather than the conventional Euclidean geometry. This means a change in the density function for homogeneous population density from $2\pi$ r dr to 4r dr (mathematical note 6.1). But average distance remains as before at $\frac{2}{3}$ of the market radius.

Maximizing profits per area results in smaller market areas than the maximization of profits per store. For in maximizing profits of a store, marginal profits at the market boundary are zero while they must equal average profits and are thus positive when maximizing profits per area. Of course, once the stores are located, each manager should just aim for the highest profits attainable.

This problem of store location or spacing can be analyzed in the conventional way with mill pricing and customers bearing a transportation cost – of themselves, not of the good.

Although this personal transportation cost is one in terms of effort, time and inconvenience, we must convert this into money terms when adding transportation cost to price of the good at the outlet, i.e. when considering mill pricing.

Fast food operators as well as department stores may also offer free home delivery (within a certain range) and this is a case of uniform pricing.

When two rival chains operate in the same industry one can observe a Hotelling phenomenon that they will locate their outlets in close proximity. An example are the coffee stores of Tchibo and Eduscho in Germany. In terms of time rather than space, the two domestic airlines of Australia, Ansett and Australian, have placed their departure times extremely close together. If a firm may choose to place a set of outlets at distances close enough to make entry unprofitable to rivals, this requires a smaller number of stores than needed to maximize profits per area, and the additional stores may be brought in later on (mathematical note 6.1).

## Public Facilities

As an example of public services with fixed demand we consider schools and their districting. For political reasons, cross-hauling of students is now practiced. We consider the older system of assigning each pupil to his/her nearest school. How large should these school districts be?

When taxpayers expenditures are the only consideration, as large as possible. To formulate a meaningful objective function, the time and effort involved in access for the school children must be expressed in a comparable unit, i.e. in money terms, as a transportation cost, even when walking rather than bussing is involved. Since the demand is inelastic and equal to a fixed fraction of population, cost minimization is equivalent to welfare maximization. For a population distribution of constant density and Manhattan metric of the street system, an elegant formula can be derived, involving only the fixed cost of each school, population density and the rate of transportation cost.

Once more the minimization of system's cost is equivalent to the minimization of cost per area or cost per pupil. For irregular population distribution the problem is analogous to the warehouse problem, to be considered below.

In a more sophisticated approach the cost of each school is not considered fixed but increasing with size at a decreasing rate due to economies of scale. For costs as a power function of size, a formula can be derived showing how the optimum size of a school and school district increases with operating cost coefficients of school, and decreases with population density and transportation cost.

It is remarkable that in a spatial context increasing returns to scale raise no real problems for an efficient allocation of resources, since transportation costs impose even stronger decreasing returns to scale.

We must also consider the case that demand for a public service decreases with distance. This distance effect may not be described well as a transportation cost to be added to the fee for the service. Rather, the distance law in operation should be brought in explicitly (Leonardi 1982). The objective function should be welfare to clients minus operating cost of the agency and welfare must be expressed in money terms. This difficulty is only partly avoided when a budget is set within whose limits the public welfare is to be maximized. For what should be the size of the budget? (Mathematical note 6.5).

## Warehouses and Fire Stations

The warehouse problem is a classical allocation problem involving indivisibilities and drew attention as the first example of a successfully computed mixed integer and linear program. A firm serving a given set of customers with given demands in discrete locations, seeks to set up an optimum number of warehouses in optimum locations, so as to minimize the sum of the fixed costs of warehouses and of the cost of transportation from these warehouses (Cooper 1954). Once the warehouses are located, the remaining problem of the best distribution of the product to customers becomes the well-known transportation problem of linear programming.

In a heuristic approach, fractional assignments of warehouses to locations may be allowed and the resulting values then be rounded up or off. Such a linear program yields insight through its efficiency prices which define an economic value for warehouses in various locations.

Public services of the emergency type (fire protection, police, ambulances) must be located sufficiently close to their customers. If this is interpreted to mean not farther than a certain critical distance, then this defines a "covering problem" (Revelle 1987). The object is finding a minimal number of service stations so located that nobody is farther than a critical distance from one service station. When costs of fire stations depend on location, this can be generalized as a cost minimization rather than a numbers minimization problem.

Once more, indivisibilities are involved requiring the solution of a mixed integer linear program. The linear program approximation yields some economic insight by placing a value on each station to be not less than its cost, and to be recovered from receipts for the services rendered. These calculations are also appropriate if such a service were privatized.

**Mathematical Note 6.1**
**The McDonald Problem**

Let R be the market radius of any branch store so that 2R is the distance between the two nearest stores.

### *Considering Demand*

Let transportation cost be per person and added to the price p of the meal per person (even in a family group) consumed at the facility. With linear cost and arbitrary demand functions profits per store are

$$(p-c)\int_0^R f(p+kr)\rho(r)\,dr - F$$

and profits per household

$$G_M = \frac{(p-c)\int_0^R f(p+kr)\rho(r)dr - F}{\int_0^R \rho(r)dr}$$

For linear demand functions this becomes

$$G_M = \frac{(p-c)\int_0^R [a-p-kr]\rho(r)dr - F}{\int_0^R \rho(r)dr}$$

$$= (p-c)[a-p-k\bar{r}(R)] - \frac{F}{M(R)} \qquad (1)$$

Determining the profit maximizing price yields

$$p = \frac{1}{2}(a+c+k\bar{r}) \qquad (2)$$

as before (3.3.14).

Substituting (2) in (1) yields profits per area as a function of market radius R

$$G(R) = \frac{k^2}{4}\left[\frac{a-c}{k} - \bar{r}\right]^2 - \frac{F}{M(R)} \qquad (3)$$

With homogenous population density in a Manhattan metric (figure 6.1)

$$\rho(r)dr = 4\rho r dr \quad \bar{r}(R) = \frac{2}{3}R \quad M(R) = 2\rho R^2 \qquad (4)$$

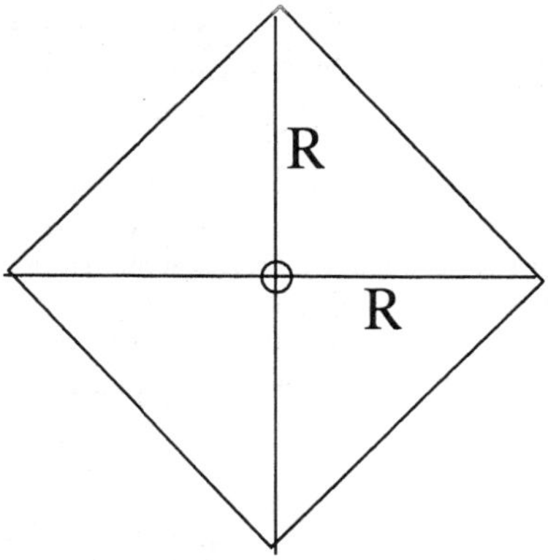


**Fig. 6.1a.** Market area of radius R in Manhattan metric

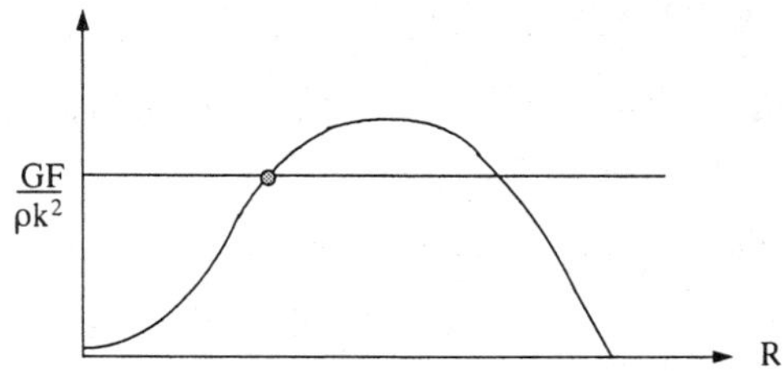


**Fig. 6.1b.** Determination of optimal market radius

$$G(R) = \frac{k^2}{4}\left[\frac{a-c}{k} - \frac{2}{3}R\right]^2 - \frac{F}{2\rho R^2} \tag{5}$$

The optimal market radius is then given by

$$0 = -\frac{2}{3}\cdot\frac{k^2}{2}\left[\frac{a-c}{k} - \frac{2}{3}R\right] + \frac{2F}{2\rho R^3}$$

$$\frac{3F}{\rho k^2} = \frac{a-c}{k}R^3 - \frac{2}{3}R^4 \tag{6}$$

Standardizing $\frac{a-c}{k} = 1 \qquad k = 1$

$$R^3 - \frac{2}{3}R^4 = \frac{3F}{\rho} \tag{7}$$

The second order conditions specifies that the left intersection be chosen. It follows that optimal spacing is an increasing function of $\frac{F}{\rho k^2}$ and a decreasing function of $\frac{a-c}{k}$

If the price of the product is set independently of considerations of market area size, the problem simplifies to

$$\operatorname*{Max}_{R}(p-c)\left[a-p-k\frac{2}{3}R\right]-\frac{F}{2\rho R^{2}}$$

yielding

$$0=-\frac{2}{3}(p-c)k+\frac{2F}{2\rho R^{2}}$$

$$R=\sqrt{\frac{3}{2}\frac{F}{\rho(r-c)k}}$$

$$R=\sqrt{\frac{3}{2}\frac{F}{\rho k(p-c)}} \tag{8}$$

a simple formula indicating that the market radius should increase with the fixed cost per store and decrease with the gross profit margin $p-c$, the customer density $\rho$ and transportation cost k. The term under the square root is dimensionless since $\rho$ has dimension $\frac{1}{\text{price}}$.

To find the spacing that will preempt the market to deter entry consider the market radius $\hat{R}$ that allows a firm to break even.

Using (5)

$$0=\frac{k^{2}\rho}{2}\left[\frac{a-c}{k}-\frac{2}{3}R\right]^{2}R^{2}-F \tag{9}$$

which corresponds to 7.1.2 below in Löschian equilibrium, mathematical note 7.1, adjusted for Manhattan metric.

This spacing is wider than that for maximizing profit per area. To show this compare standardized versions

$$\left(1-\frac{2}{3}R\right)^{2}R^{2}=\frac{2F}{\rho} \tag{9a}$$

and

$$R^{3}-\frac{2}{3}R^{4}=\frac{3F}{\rho} \tag{6a}$$

Rewrite

$$\frac{6F}{\rho}=3\left(1-\frac{2}{3}R\right)^{2}R^{2}=\phi_{1}(R) \qquad \text{(say)}$$

$$\frac{6F}{\rho}=2R^{3}-\frac{4}{3}R^{4}=\phi_{2}(R) \qquad \text{(say)}$$

and consider

$$\frac{\phi_1 - \phi_2}{R^2} = 3 - 2R - 2R + \frac{4}{3}R^2$$

$$= \frac{4}{3}\left(R - \frac{3}{2}\right)^2 + \frac{9}{4} - \frac{9}{16} > 0$$

**Fig. 6.1c.** Break-even versus optimal market radius

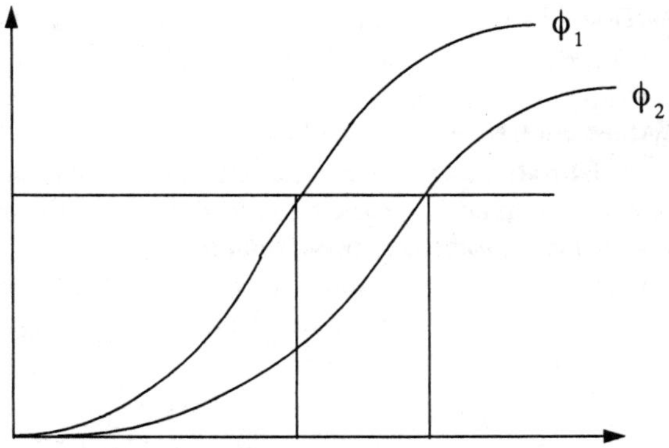


Since

$$\phi_1(R) > \phi_2(R)$$

the break-even market size is less than that for maximum profits per area, as might be expected.

But the question is whether a new outlet by a rival located between four outlets of the existing firm can break even. The contending outlet would have a market radius of $\frac{\sqrt{2}}{2}R$ (figure 6.1d).

**Fig. 6.1d.** Entry of rival

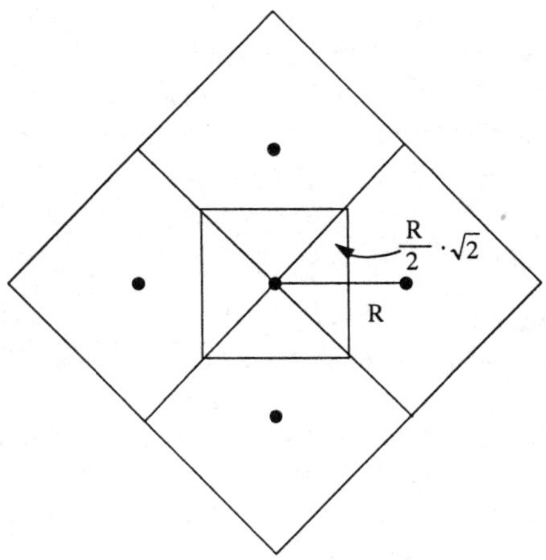

Now

$$\phi_1\left(\frac{R}{\sqrt{2}}\right)=\left(1-\frac{\sqrt{2}}{3}R\right)\frac{R^2}{2}\gtrless R^3-\frac{2}{3}R^4=\phi_2(R)=\frac{6F}{\rho}$$

Hence a firm locating its outlets at distances that maximize profits per area is vulnerable to entry by rivals who are content to just making a profit. Reality shows indeed examples of co-existing chains, particularly in the fast food industry, e.g. McDonald's, Burger King, Wendy's, Kentucky Fried Chicken and Pizza Hut, to name but a few.

**Mathematical Note 6.2**
**Home Delivery**

Profits per household to be served are

$$G_U=\frac{f(p)}{M(R)}\int_0^R[p-c-kr]\rho(r)dr-\frac{F}{M(R)} \tag{1}$$

$$=f(p)[p-c-k\bar{r}(R)]-\frac{F}{M(R)}$$

For uniform population density and Manhattan metric

$$G_U=f(p)\left[p-c-\frac{2}{3}kR\right]-\frac{F}{2\rho R^2} \tag{2}$$

For given prices the optimal market radius is determined by

$$0=-\frac{2}{3}f(p)k+\frac{2F}{2\rho R^3}$$

$$R=\sqrt[3]{\frac{\frac{3}{2}F}{\rho kf(p)}} \tag{3}$$

which corresponds to (6.1.8) with demand f(p) in lieu of a fixed profit margin p-c.

Observe that both $\rho$ and f(p) are dimensionless here. A comparison of (6.1.8) and (3) shows that market areas for free delivery are larger when demand per household is numerically smaller than the profit margin in mill pricing. When both price and market size are chosen to maximize profits per area and demand functions are linear, then they are equal for home delivery or consumption at the outlet.

**Mathematical Note 6.3**
**School Districting in a Homogenous Urban Area**

In the school location or districting problem the demand is fixed, say proportional to population and the object is to minimize transportation cost of children plus fixed costs of schools. For a school population of constant density $\rho$ in an area with Manhattan geometry the cost per pupil to be minimized is then

$$\frac{F + k\bar{r}M}{M} = \frac{F}{M(R)} + k\bar{r}(R) \tag{5}$$

For Manhattan metric

$$\bar{r} = \frac{2}{3}R \tag{6}$$

$$M = 2\rho R^2 \tag{7}$$

the minimum problem is solved by

$$0 = \frac{2}{3}k - \frac{2F}{2\rho R^3}$$

$$R = \sqrt[3]{\frac{3}{2}\frac{F}{\rho k}} \tag{8}$$

a simple formula. Spacing is closer the greater the population density and transportation cost and the smaller the fixed cost.

**Mathematical Note 6.4**
**Increasing Returns in Schools and Optimal District Size**

Let Q be the number of pupils and costs of the school

$$bQ^{\beta}$$

Increasing returns mean decreasing unit costs, i.e.

$$0 < \beta < 1 \tag{9}$$

A school district of radius R and size $2R^2$ (in Manhattan geometry) in a homogeneous urban area produces a student population $2\rho R^2$ (Here as in 6.3 before $\rho$ denotes not household density, as usual, but student density, that is, students per area). Minimizing costs per student

$$\underset{R}{\text{Min}} \frac{b\left[2\rho R^2\right]^{\beta} + \rho k \int_0^R r4rdr}{2\rho R^2}$$

$$= \underset{R}{\text{Min}}\, b\left[2\rho R^2\right]^{\beta-1} + \frac{2}{3}kR \tag{10}$$

yielding

$$0 = (\beta - 1)b(2\rho)^{\beta-1}R^{2\beta-3} + \frac{2}{3}k$$

$$R = \left[\frac{3b(1-\beta)(2\rho)^{\beta-1}}{2k}\right]^{\frac{1}{3-2\beta}} \tag{11}$$

for $\beta < \frac{3}{2}$ a power function increasing in the school cost b and the exponent β and decreasing with respect to transportation cost and student density.

**Mathematical Note 6.5**
**Locating Services When Demand Decreases With Distance**

u = utility (value) of service per client
$\gamma(r)$ = demand for services as a function of distance r
F = fixed cost, the only cost of the service

The object is to maximize the utility of services rendered minus cost per area population

$$\underset{R}{\text{Max}} \frac{u\int_0^R \gamma(r)\rho(r)dr - F}{M(R)} \tag{1}$$

For Manhattan geometry and constant population density

$$\underset{R}{\text{Max}} \frac{u\rho\int_0^R \gamma(r)4rdr - F}{2\rho R^2}$$

$$= \frac{2u\int_0^R \gamma(r)rdr}{R^2} - \frac{F}{2\rho R^2} \tag{2}$$

This is equivalent to

$$\text{Max} \frac{\int_0^R \gamma(r)rdr}{R^2} - \frac{F}{4\rho u R^2} \tag{3}$$

Setting the derivative with respect to R zero

$$0 = \frac{R^3\gamma(R) - 2R\int_0^R \gamma(r)rdr}{R^4} + \frac{F}{2\rho u R^3}$$

$$-R^2\gamma(R) + 2\int_0^R \gamma(r)rdr = \frac{F}{2\rho u} \tag{4}$$

$$H(R) = \frac{F}{2\rho u} \qquad \text{(say)} \tag{5}$$

**Fig. 6.5.** Determination of service area

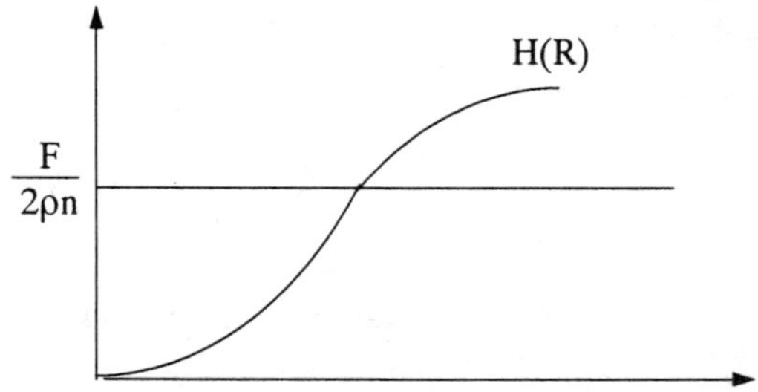


Now H(R) is an increasing function of R

$$\frac{dH}{dR} = -2R\gamma(R) - R^2\gamma'(R) + 2\gamma(R)R$$

$$= -R^2\gamma'(R) > 0$$

since interaction decreases with distance $\gamma' < 0$.

It follows from (5) that the optimal service area increases with (fixed) cost F and decreases with population density and the utility of service, as one would expect.

For exponential distance effects and standardized distance

$$\gamma(r) = e^{-r} \tag{6}$$

$$H(r) = -r^2e^{-r} + 2\int_0^r e^{-t}tdt$$

$$= 2 - \left(r^2 + 2r + 2\right)e^{-r}$$

$$H(0) = 0 \qquad H'(r) = r^2e^{-r} > 0$$

## Mathematical Note 6.6
## The Warehousing Problem

This problem attained some prominence in the operations research literature of the 50s. Define

$x_i = 0,1$ presence of a warehouse in point location i
$x_{ij} \geq 0$ flow of the commodity from warehouse i to market j
$q_j$ (fixed) demand in market j
$k_{ij}$ transportation cost per unit from i to j
F the (fixed) cost of a warehouse
c the capacity of a warehouse.

The mixed integer program, linear program is

$$\min_{\substack{x_i = 0,1 \\ x_{ij} \geq 0}} F\sum_i x_i + \sum_{ij} k_{ij}x_{ij} \tag{1}$$

subject to

$$\sum_i x_{ij} \geq q_j \tag{2}$$

$$\sum_j x_{ij} \leq cx_i \tag{3}$$

In a first approach the capacity c may be assumed large enough to impose no actual constraint on the total level of flows $\sum_j x_{ij}$.

Once the $x_i$ are chosen this becomes a linear programming transportation problem which is easily solved. The difficult part is the determination of the warehouse locations $x_i$.

In a linear programming formulation the condition $x_i = 0,1$ is replaced by the weaker one

$$0 \leq x_i \leq 1 \tag{4}$$

admitting as it were, fractional warehouses, which will be either rounded off, i.e. eliminated, or rounded up, i.e. placed. Now the capacity c must be assumed to impose effective limits. The Lagrange function of this linear program is then

$$L = -F\sum_i x_i - \sum_{ij} k_{ij}x_{ij} + \sum_j \lambda_j\left(\sum_i x_{ij} - q_j\right)$$

$$+\sum_i \mu_i\left(cx_i - \sum_j x_{ij}\right) + \sum_i v_i(1 - x_i) \tag{5}$$

The efficiency conditions are

$$\begin{Bmatrix} x_i^* = 0 \\ 0 \leq x_i \leq 1 \\ x_i = 1 \end{Bmatrix} \Leftrightarrow c\mu_i \begin{Bmatrix} < \\ = \\ > \end{Bmatrix} F \tag{6}$$

$$x_{ij}\begin{Bmatrix} = \\ \geq \end{Bmatrix} 0 \Leftrightarrow \lambda_j - \mu_i \begin{Bmatrix} < \\ = \end{Bmatrix} k_{ij} \tag{7}$$

Here $\mu_i$ may be interpreted as the value or price of the good in warehouse i, and $\lambda_j$ its value or sales price in market j. Condition (7) states that transportation costs must be recovered and (6) that the warehouse costs F must be earned.

**Mathematical Note 6.7**
**The Covering Problem**

| | |
|---|---|
| $i = 1, \ldots m$ | potential locations of fire stations |
| $j = 1, \ldots n$ | residences, to be protected |
| $S_i$ | set of residences j within reach of station at i |

| | |
|---|---|
| $F_i$ | cost of station at i |
| c | engine capacity of a fire station |
| $q_j$ | demand for fire engines in j |
| $x_i = 0,1$ | placement of fire stations in i |
| $x_{ij} \geq 0$ | trips of fire engines from i to j- |

Objective

$$\begin{aligned} &\text{Min} \quad \sum_i F_i x_i \\ &x_i = 0,1 \\ &x_{ij} \geq 0 \end{aligned} \tag{1}$$

subject to

$$\sum_j x_{ij} \leq c \cdot x_i \tag{2}$$

$$\sum_i x_{ij} \geq q_j \tag{3}$$

Relaxing the constraint $x_i = 0,1$ to

$$0 \leq x_i \leq 1$$

generates a linear program with a Lagrangean

$$L = -\sum_i F_i x_i + \sum_i \mu_i \left( c x_i - \sum_j x_{ij} \right) + \sum_{j \in S_i} \lambda_j \left( \sum_i x_{ij} - q_j \right) \tag{4}$$

and efficiency conditions

$$x_i \begin{Bmatrix} = \\ \geq \end{Bmatrix} 0 \Leftrightarrow c\mu_i \begin{Bmatrix} < \\ = \end{Bmatrix} F_i \tag{5}$$

$$x_{ij} \begin{Bmatrix} = \\ \geq \end{Bmatrix} 0 \Leftrightarrow \lambda_j \begin{Bmatrix} < \\ = \end{Bmatrix} \mu_i \tag{6}$$

$$\mu_i \begin{Bmatrix} = \\ \geq \end{Bmatrix} 0 \Leftrightarrow \sum_j x_{ij} \begin{Bmatrix} < \\ = \end{Bmatrix} c \tag{7}$$

$$\lambda_j \begin{Bmatrix} = \\ \geq \end{Bmatrix} 0 \Leftrightarrow \sum_i x_{ij} \begin{Bmatrix} > \\ = \end{Bmatrix} q_j \tag{8}$$

From (6) and (7) it follows that when $q_j>0$

$$\lambda_j = \min_{j \in s_i} \mu_i \tag{9}$$

The cheapest station within reach is allocated to any residence. (5) says that fire stations be located only when their value $c \cdot \mu_i$ covers their cost $F_i$. This value is earned from the value of services delivered to residences j. For

$$\sum_j x_{ij}\lambda_j = \mu_i \sum_j x_{ij} = \mu_i c \tag{10}$$

using (6) and (7). The linear program will choose too many stations at fractional levels. These must then be rounded up or rounded off to obtain an approximation to the best integer solution.

## References

Church F, ReVelle C (1973) The Maximal Covering Location Problem. Papers of the Regional Science Association 32:101–118

Cooper L (1954) The Warehouse Problem. Operations Research

Leonardi G (1982) A Multiactivity Location Model with Accessibility and Congestion – Sensitive Demand. Sistemi Urbani 3:267–310

Lovell M (1970) Product Differentiation and Spatial Structure. Western Economic Journal 8:120–143

Revelle C (1987) Urban Public Facility Location. In: Mills ES (ed) Handbook of Regional and Urban Economics, 2 Urban Economics. North Holland, Amsterdam

Schuler RE, Holahan WL (1977) Optimal Size and Spacing of Public Facilities in Metropolitan Areas. Papers of the Regional Science Association, August, p 137–156

Thisse J-F, Zoller HG (eds) (1983) Locational Analysis of Public Facilities. North Holland, Amsterdam

Toregas C, Swain R, ReVelle C, Bergman L (1971) The Location of Emergency Facilities. Operations Research 19, 6 (October):1363–1373

Wagner JL, Falkson LM (1975) The Optimal Nodal Location of Public Facilities. Geographical Analysis 7, 1 (January):68–83

Walker W (1974) Using the Set-covering Problem to Assigne Fire Companies in Fire Houses. Operations Research 22, 2 (March–April):275–277

CHAPTER 7

# Dispersion

## Löschian Equilibrium

Complete dispersion of an economic activity means that it is performed in every household. For activities with increasing returns to scale, this is not a stable equilibrium. Rather, as Lösch describes it for the example of beer brewing, some households will specialize in it and exchange the product for other goods and services at an advantage, over producing these other goods and services itself. This decrease of dispersion can also be viewed as a "break of symmetry" (I Stewart 1995).

In discussing spatial equilibrium of a market oriented dispersed industry, the location of any particular plant is not the subject. Rather one wants to know the number of plants per area, the sizes of market areas and the distance at which plants are spaced, all of which are equivalent questions. How many households will the brewer chose to supply? This is the question of profit maximizing market areas, studied in section 3.4. Initially the market will be a circle. Under mill and uniform pricing its radius is chosen as

$$\frac{3}{4}\frac{a-c}{k}$$

and under discriminatory pricing at the maximum size $\frac{a-c}{k}$.

This remains so as long as the producers of this product are few enough so as to not encroach on each other's territory.

When free entry causes market areas to become contiguous, firms are no longer in a profit maximizing monopoly position but, facing approximately six neighbors, enter into oligopoly. Of the various pricing strategies that might then be adopted, one of price leadership aimed at continued monopolistic pricing is the simplest to chose. This is the Löschian assumption. Free entry will now reduce market areas.

We postpone a discussion of the shape of these and assume that market areas are still approximately circular. Now profits, under whatever pricing strategy must decrease with market radius. R, measured as the radius of the largest inscribed circle, which is one half the distance to nearest neighbors. When R has reached the level of vanishing profits, an equilibrium is established, that in non-spatial theory is known as that for monopolistic competition with free entry. It was recognized as its spatial equivalent by Palander (1935) and Lösch (1941). The equilibrium radius is an increasing function of fixed cost and a decreasing function of transportation cost. For linear demand functions explicit formulas can be given (mathematical note 7.1).

What happens to prices as market areas shrink? Mill price in which average distance to customers enters negatively will rise but uniform prices decline as markets shrinks, but average prices paid by consumers, being an increasing function of the market radius, always fall when markets become smaller. Whether market sizes reached in equilibrium are also those that create maximum welfare is examined in mathematical notes 15.5 and 15.6.

Ever since Lösch (1940) location theorists have been fascinated by the image of hexagonal market areas and their apparent perfection. Even as a purely theoretical construct their validity is restricted, however. Consider an industry whose representative firm just manages to break even, i.e. cover its fixed cost by charging monopoly prices and serving a circular market area at the optimal radius $\frac{3}{4}\frac{a-c}{k}$ so that plants must have a minimum distance of $\frac{3}{2}\frac{a-c}{k}$.

Customers located between three adjacent circles will remain unserved. For hexagons of the same radius, in which average distance from the center is larger, (table 2.2), would not yield a sufficient profit to sustain them. As we lower fixed cost, then markets are allowed to shrink but the squeezing in of additional firms results in market areas being flattened into rounded hexagons that still leave corner areas unserved (figure 7.1). The fixed cost levels yielding hexagons and those permitting only circles can be calculated (mathematical note 7.2). That hexagons are not the only possible outcome of a market area squeezing by free entry was first discovered by Mills and Lav (1964).

Examine now the shape of market areas. Through entry of others, a given firms' market and those of its six neighbors are separated by line segments as borders which at first leave some circular segments in between, which are borders of certain corners of unserved territory (figure 7.1).

Depending on the size of fixed cost, free entry equilibrium can be reached when market areas have this shape of flattened circles or rounded hexagons (Beckmann 1971). If entry and squeezing continue then eventually full fledged hexagons will emerge as equilibrium market areas (Mills and Lav 1964).

**Fig. 7.0.** Rounded hexagons as market areas

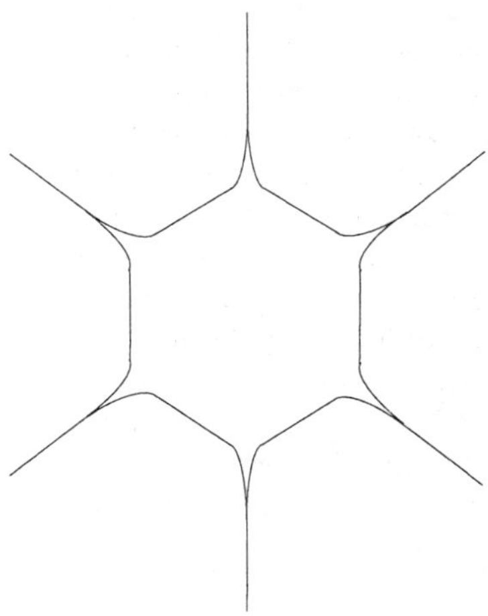

One can calculate the range of fixed cost levels for which hexagons or the approximate hexagons are the equilibrium solution (mathematical note 7.2).

It can be argued, of course, that in reality the condition of exactly vanishing profits is too strict. Rather that it would be enough that new firms which would enter at the corners of market areas between three adjacent firms, which would be the most promising ones, do no longer yield an expected profit and this would create a range of equilibrium market radii rather than a unique level (Eaton and Lipsey 1976).

## Uniform Pricing: Bertrand Oligopoly

When uniform pricing is considered, the case for price leadership is less convincing. Rather one would expect Bertrand oligopoly to prevail in view of the obvious possibility of increasing market size (at least initially) by setting a price that undercuts those of neighbors'. An equilibrium can then be reached where the existing price separates market areas since it is no longer profitable to ship into neighbors territories. There exists a density of firms for which this equilibrium among market areas also permits all firms just to break even.

These equilibrium market areas must be circular and are thus different from those in Löschian equilibrium. The market radius is larger than the radius of the inscribed circle for market areas in Löschian equilibrium so that fewer firms survive (mathematical note 7.3).

The market areas emerging with perfectly discriminating pricing are similar to those of Lösch with mill pricing but greater profitability means smaller equilibrium market radii, hence the survival of more firms. This is the realization of Lösch's principle that spatial equilibrium of a dispersed industry "maximizes the number of independent existences" (Lösch 1956).

## Other Pricing Strategies

Other pricing strategies must also be studied. These include Cournot pricing where mill prices, rather than quantities, of neighboring firms are accepted as unchanging and the own price is adapted accordingly.

Of interest is also pricing under consistent conjectural variations. Their complexity has limited the analysis to one dimensional scenarios. A comparison of results with those for two-dimensional markets is thus difficult.

### Mathematical Note 7.1
### The Lösch Model for Circular Markets

Profits as a function of market radius were derived in mathematical note 3.4 as

$$G_M(R) = \frac{k^2}{4}\left[\frac{a-c}{k} - \bar{r}(R)\right]^2 M(R) - F \tag{1}$$

for linear demand functions q = a - p and profit maximizing mill pricing.

With uniform population density $\rho(r) = 2\pi\rho r$ this becomes a quartic function

$$G_M(R) = \frac{k^2}{4}\left[\frac{a-c}{k} - \frac{2}{3}R\right]^2 \pi\rho R^2 - F \tag{2}$$

It is an increasing function of R since, dropping $\frac{k^2}{4}\pi\rho$

$$\frac{dG_M}{dR} \approx -\frac{4}{3}\cdot\left(\frac{a-c}{k} - \frac{2}{3}R\right)R^2 + \left(\frac{a-c}{k} - \frac{2}{3}R\right)^2 2R$$

$$\sim +\frac{4}{9}R^2 + \left(\frac{a-c}{k} - \frac{2}{3}R\right)$$

$$\sim -\frac{4}{3}R + \frac{a-c}{k} > 0$$

since

$$R \leq \frac{3}{4}\frac{a-c}{k},$$

the optimal radius

Now free entry decreases R until profits vanish (figure 7.1).

$$\left[\frac{a-c}{k} - \frac{2}{3}\hat{R}\right]^2 \hat{R}^2 = \frac{4F}{\pi\rho k^2} \tag{3}$$

Notice that the right hand expression is dimensionless since (for linear demand functions) $\rho$ has dimension $\frac{1}{\text{price}}$. Since the left hand side of (3) is an increasing function of R, the equilibrium market radius $\hat{R}$ increases with $\frac{F}{\rho k^2}$. Fixed cost F has an increasing, customer density $\rho$ a decreasing effect on the size of market areas and the spacing of firms.

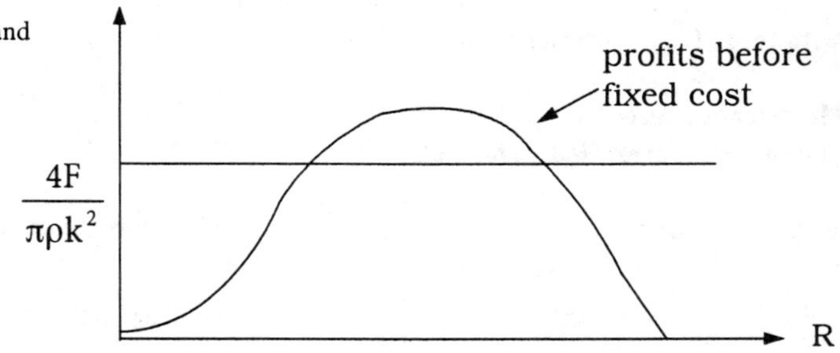


**Fig. 7.1.** Gross profits and distance

To study the effects of a - c and k perform implicit differentiation of

$$\left[a - c - k\hat{R}\right]^2 \hat{R}^2 \equiv \phi(R, a - c, k) = \frac{4F}{\pi\rho}$$

$$\text{sign}\,\frac{\partial \hat{R}}{\partial(a - c)} = \text{-sign}\,\frac{\phi_{a-c}}{\phi_R} - \frac{+}{+} = -$$

$$\text{sign}\,\frac{\partial \hat{R}}{\partial k} = -\,\text{sign}\,\frac{\phi_k}{\phi_R} = -\frac{-}{+} = +$$

so that the equilibrium market radius falls with the spread a-c between demand intercept and unit cost and rises with transportation cost. In particular transportation cost has a negative effect on the number of firms that can survive.

Now (3) can be solved explicitly for $\hat{R}$.
Abbreviating

$$\frac{a - c}{k} = m \qquad \frac{4F}{\pi\rho k^2} = n^2$$

and taking the square root in (3)

$$\hat{R} \cdot \left(m - \frac{2}{3}\hat{R}\right) = n$$

$$\hat{R}^2 - \frac{3}{2}m\hat{R} = -\frac{3}{2}n$$

$$\left(\hat{R} - \frac{3}{4}m\right)^2 = \frac{9}{16}m^2 - \frac{3}{2}n$$

$$\hat{R} = \frac{3}{4}m - \sqrt{\frac{9}{16}m^2 - \frac{3}{2}n}$$

where the – sign applies since

$$\hat{R} < \frac{3}{4}m$$

the monopoly market radius. Thus

$$\hat{R} = \frac{3}{4}m\left[1 - \sqrt{1 - \frac{8}{3}\frac{n}{m^2}}\right]$$

$$= \frac{3}{4}\frac{a - c}{k}\left[1 - \sqrt{1 - \frac{16k}{3}\frac{\sqrt{\dfrac{F}{\pi\rho}}}{(a - c)^2}}\right] \tag{4}$$

confirming that the equilibrium market radius must decrease with transportation cost k.

## Mathematical Note 7.2
## Mills-Lav Effect

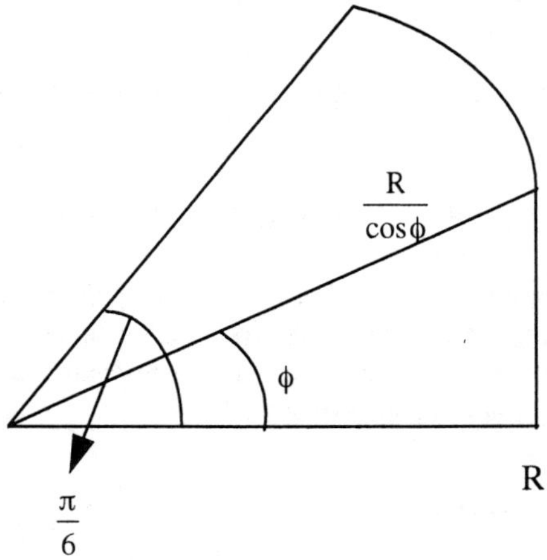


**Fig. 7.2a.** A segment of $\frac{1}{12}$ of the market area in the shape of a flattened circle

The profit function under optimal mill pricing

$$p_M = \frac{1}{2}(a + c - k\bar{r}) \tag{1}$$

may be constructed by looking at a slice of $\frac{1}{12}$ the market area, enclosed in an angle $\frac{2\pi}{12} = \frac{\pi}{6}$ (figure 7.2).

$$G_M = \rho(p_M - c) \cdot 12 \left[ \left(\frac{\pi}{6} - \phi\right) \int_0^{\frac{R}{\cos\phi}} [a - p_M - kr] r \, dr \right.$$

$$\left. + \int_0^{\phi} d\psi \int_0^{\frac{R}{\cos\psi}} [a - p_M - kr] r \, dr \right] - F \tag{2}$$

Shape is determined by the optimal angle $\phi$, given the market radius R

$$\frac{\partial G_M}{\partial \phi} = 0 = - \int_0^{\frac{R}{\cos\phi}} \quad + \quad \int_0^{\frac{R}{\cos\phi}}$$

$$+ \left(\frac{\pi}{6} - \phi\right) \cdot \left[ a - p_M - k \frac{R}{\cos\phi} \right] \cdot \frac{R \sin\phi}{\cos^2\phi} + \frac{\partial G}{\partial p_M} \cdot \frac{dp_M}{d\phi}$$

But

$$\frac{\partial G}{\partial p_M} = 0$$

by profit maximization
Hence

$$a - p_M - k\frac{R}{\cos\phi} = 0$$

Recall (1)
so that

$$\frac{2R}{\cos\phi} = \frac{a-c}{k} + \bar{r} \tag{3}$$

For $\phi = 0$ we have circles and so

$$\bar{r} = \frac{2}{3}R$$

yielding

$$R = \frac{3}{4}\frac{a-c}{k}$$

Standardizing

$$R = \frac{3}{4},$$

the monopoly market radius.
For hexagons

$$\phi = \frac{\pi}{6}$$

$$\cos\phi = \frac{\sqrt{3}}{2} \quad \bar{r} = .702R$$

$$\frac{4R}{\sqrt{3}} = 1 + .702R$$

$$R = \frac{1}{\frac{4}{\sqrt{3}} - .702} = .622$$

The range of market radii in which market areas are flattened circles or rounded hexagons is thus

$$.622 \le R \le \frac{3}{4} = .75 \tag{4}$$

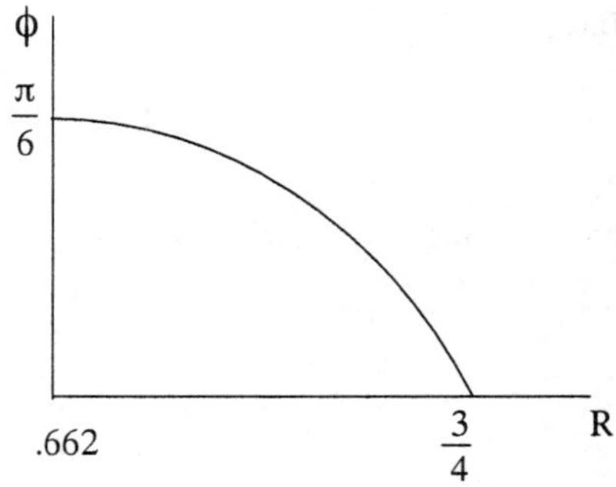


**Fig. 7.2b.** Angle $\phi$ of the straight line segments of the boundary as a function of radius R (Calculations courtesy of T Puu)

In figure 7.2b the angle $\phi$ of the straight line border segments is shown as a function of the equilibrium market radius R, using (3).

Given $\phi(R)$ this equilibrium market radius is determined by setting the profit function (2) equal to zero. *Ceteris paribus* this equilibrium market radius is larger the larger is fixed cost F.

### Mathematical Note 7.3
### Bertrand Oligopoly with Uniform Prices

Price cutting under Bertrand oligopoly means that marginal costs are just covered at the market boundary

$$p_U = c + kR$$

In equilibrium market areas are separated – but touching – and fixed costs are just covered

$$f(c+kR)\cdot\int_0^R (p-c-kr)\rho(r)dr = F$$

$$f(c+kR)\cdot\int_0^R (R-r)\rho(r)dr = \frac{F}{k}$$

$$f(c+kR)\cdot[R-\bar{r}]M(R) = \frac{F}{k} \tag{2}$$

For linear demand and constant population density

$$\left(\frac{a-c}{k} - R\right)R^3 = \frac{3F}{k^2\pi\rho} \tag{3}$$

or standardized

$$(1-R)R^3 = \frac{3F}{\pi\rho} \tag{4}$$

To compare this with Löschian market equilibrium

$$\left(1-\frac{2}{3}R\right)^2 R^2 = \frac{4F}{\pi\rho} \tag{5}$$

equalize the right hand sides of (4) and (5)

$$\frac{12F}{\pi\rho} = 4R^3 - 4R^4 = \phi_U(R) \qquad \text{(say)}$$

$$\frac{12F}{\pi\rho} = 3\left(1-\frac{4}{3}R+\frac{4}{9}R^2\right)R^2 = \phi_M(R) \qquad \text{(say)}$$

Consider

$$\frac{\phi_U - \phi_M}{R^2} = -\frac{16}{3}R^2 + 8R - 3$$

$$= -\frac{16}{3}\left(R-\frac{3}{4}\right)^2 < 0$$

Since $\phi_U < \phi_M$ for all R the equilibrium market sizes determined by the intersection of $\phi(R)$ with the constant line $\frac{12F}{\pi\rho}$ (figure 7.3) are

$$R_M < R_U$$

The less profitable Bertrand oligopoly under uniform pricing results thus in larger market areas for surviving firms. Notice that uniform pricing leaves certain areas between three adjacent circular markets unserved. They represent a fraction of

$$1-\frac{\pi}{2\sqrt{3}}$$

or 9.31%.

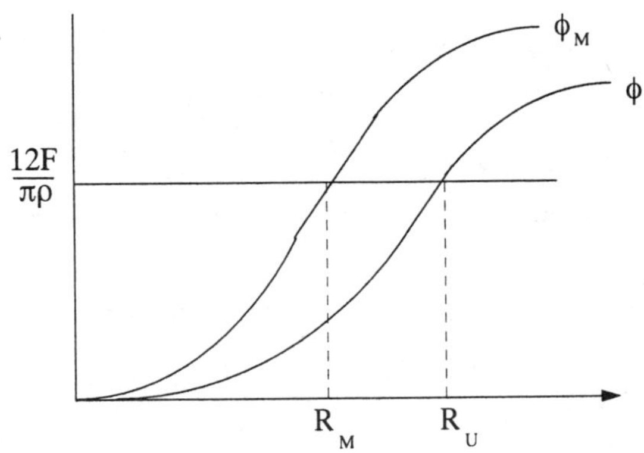


**Fig. 7.3.** Löschian equilibrium versus Bertrand oligopoly

## Mathematical Note 7.4
## Cournot Oligopoly with Mill Pricing

In a spatial context, Cournot oligopoly is possible with prices as strategic variables, not just with quantities. This is because prices affect market areas. How the extent of a market area depends on pricing is brought into sharp focus in mill pricing. To keep the problem manageable assume a linearly extended market and consider a firm i = 2 whose neighbor i = 1 is to the left at a distance $D_{12}$, and i = 3 is to the right at distance $D_{23}$. The firm's market radius to the left is then

$$\frac{D_{12}}{2}+\frac{p_1-p_2}{2k} \tag{1}$$

and to the right

$$\frac{D_{23}}{2}+\frac{p_3-p_2}{2k} \tag{2}$$

since

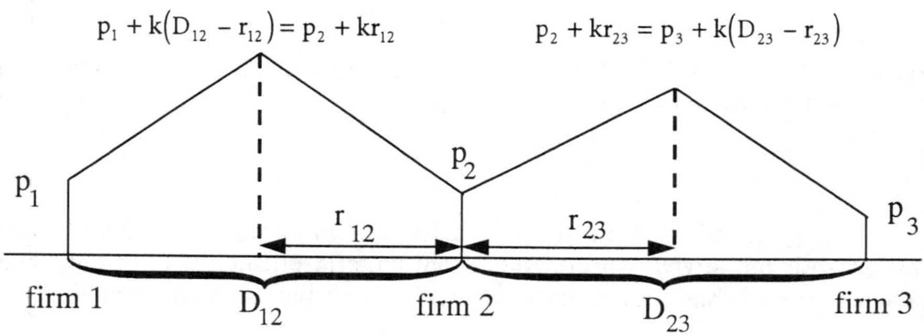


**Fig. 7.4.** Market areas under mill pricing

For simplicity write $\frac{D_{ij}}{2}=R_{ij}=\text{half distance}$

Let demand be rectangular, say

$$q(p)=\begin{cases}0 & p>a\\ 1 & p\le a\end{cases} \tag{3}$$

and assume

$$a\ge c+\frac{3}{2}kD \tag{4}$$

Then firm 2's profit is

$$G_2=(p_2-c_2)\,\text{market area}-F$$

or

$$G_2 = (p_2 - c_2)\left[R_{12} + R_{23} + \frac{1}{2k}(p_1 + p_3 - 2p_2)\right] - F = G_2(p_1, p_2, p_3) \tag{5}$$

a quadratic function in $p_1$, $p_2$, and $p_3$.

Profit maximization by firm 2 yields

$$0 = \left[R_{12} + R_{23} + \frac{1}{2}\frac{(p_1 + p_3 - 2p_2)}{k}\right] - \frac{(p_2 - c_2)}{k} \tag{6}$$

$$p_2 = \frac{c_2}{2} + \frac{p_1 + p_3}{4} + \frac{k}{2}(R_{12} + R_{23})$$

Thus firms respond positively to neighbor's prices. When unit costs $c_i$ and the market radii $R_{ij} = \frac{D}{2}$ are the same for firms i , prices are equal and the Cournot equilibrium price turns out to be

$$p = c + kD \tag{7}$$

with

$$p + k\frac{D}{2} \leq a \tag{8}$$

assured by assumption (4).

For close spacing Cournot oligopoly brings prices down from the monopoly level

$$p_i = a - kR \tag{9}$$

for rectangular demand to one that is the sum of unit production cost c and transportation cost kD over the distance D to nearest neighbors.

for

$$c + 2kR < a - kR$$

when

$$R < \frac{1}{3}\frac{a - c}{k}.$$

Increasing the number of firms and thereby moving them closer together will according to (7) lower prices at the mill and even more average delivered prices at the buyer's site.

Under free entry the profits of each firm

$$\begin{aligned} G &= (p - c) \cdot D - F \\ &= kD^2 - F \end{aligned}$$

are driven to zero so that the equilibrium spacing (in a linear economy for rectangular demand) is simply

$$\hat{D} = \sqrt{\frac{F}{k}} \tag{10}$$

increasing with fixed cost F, decreasing with transportation cost k and independent of marginal production cost c, as one should expect.

**Mathematical Note 7.5**
**Consistent Conjectural Variation: Linear Market, Rectangular Demand, Mill Pricing**

Let firms be spaced equally and R be the range of the market area in either direction. Profits are

$$G = 2(p - c)R(p, p_1) \tag{1}$$

for a market radius determined by (figure 7.4)

$$p + kR = p_1 + k(D - R)$$

$$2R = D + \frac{p_1 - p}{k} \tag{2}$$

where p is the mill price of the two neighbors.

The optimum price is found from

$$0 = 2R(p, p_1) + 2(p - c)\left[\frac{\partial 2R}{\partial p} + \frac{\partial 2R}{\partial p_1}\frac{dp_1}{dp}\right]$$

$$= 2R(p, p_1) + (p - c)\left[-\frac{1}{k} + \frac{\beta}{k}\right] \tag{3}$$

assuming a conjectural variation

$$\frac{dp_1}{dp} = \beta \tag{4}$$

Substituting (2)

$$0 = kD + p_1 - p + (p - c)(\beta - 1)$$

implying

$$(2 - \beta)p = p_1 + \text{const.} \tag{3}$$

Thus the conjectural variation for the firm considered is

$$\frac{dp}{dp_1} = \frac{1}{2 - \beta} \tag{5}$$

Assuming it to be the same for all firms

$$\beta = \frac{1}{2 - \beta}$$

implying

$$(\beta - 1)^2 = 0 \tag{4}$$

so that consistent conjectural variation requires

$$\beta = 1$$

This means that prices are matched justifying the Löschian assumption of monopoly pricing under price leadership.

### Mathematical Note 7.6
### Contesting Duopoly with Price Discrimination

The profit maximizing discriminating price was seen to be

$$p(r) = \frac{a-c}{2} + \frac{k}{2} r$$

when demand functions are linear and prices can be set freely. Two firms practicing this type of pricing and locating at distance D, $D > 2\,R_{max}$ where

$$R_{max} = \frac{a-c}{k}$$

may undercut each other's discriminatory price however by charging their own marginal cost price (figure 7.5.)

In the contested zone market areas are still separated at the mid point between the two firms, but paradoxically prices fall here with distance. But: the contesting firm does not gain any profit from this and exposes itself to retaliation. Also such bizarre pricing patterns draw attention to the fact of price discrimination, which is in fact illegal in many countries.

Exceptions to this appear to occur in setting air fares. But the lower fares to more distant points often reflect the economies of greater load factors for trips between major traffic generating locations. It can be observed occasionally that products are priced higher at the firm's own location than in some location more accessible to rival suppliers.

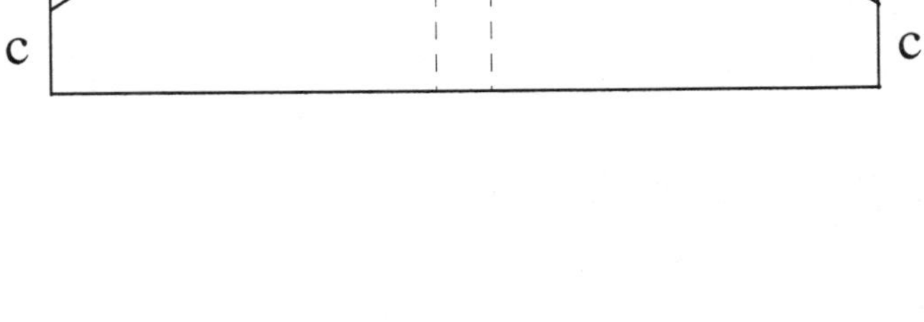


**Fig. 7.5.** Possible price pattern under discriminatory pricing

## References

Beckmann MJ (1973) Spatial Oligopoly as a Noncooperative Game. International Journal of Game Theory 2, 6:263–268

Beckmann MJ (1972) Spatial Oligopoly Revisted: An Examination of Some Strategies in Mill Pricing in One and Two Dimensional Markets. Papers and Proceedings of Regional Science Association 17, Northeast Regional Science Review 1:1–20

Beckmann MJ (1971) Equilibrium Versus Optimum: Spacing of Firms and Patents of Market Areas. Northeast Regional Science Review, p 1–20

Beckmann MJ (1989) On the Shape and Size of Market Areas. The Annals of Regional Science 23:81–91

Bos HC (1965) Spatial Dispersion of Economic Activity. Rotterdam University Press, Rotterdam

Böventer von E (1962) Theorie des räumlichen Gleichgewichts. Mohr, Tübingen

Capozza DR, Van Order R (1978) A Generalized Model of Spatial Competition. American Economic Review 68:896–908

Capozza DR, Van Order R (1989) Spatial Competition With Consistent Conjectures. Journal of Regional Science 29:1–13

Denike KG, Parr JK (1970) Production in Space, Spatial Competition and Restricted Entry. Journal of Regional Science, 10, April, p 49–63

Eaton B (1972) Spatial Competition Revisited. Canadian Journal of Economics 5:268–277

Eaton BC, Lipsey RG (1976) The Non-Uniqueness of Equilibrium in the Löschian Location Model. American Economic Review, March, p 77–93

Hartwick J (1972) Lösch Theorems on Hexagonal Market Areas. Journal of Regional Science 13:213–221

Holahan WL (1974) Hospital Behavior in a Spatial Economy, Ph. D. Thesis. Brown University

Lösch R (1954) The Economics of Location. Yale University Press, New Haven

Mills ES, Lav MR (1964) A Model of Market Areas With Free Entry. Journal of Political Economy 73:278–288

Paelinck JHP, Ancot JP, Kuiper JK (1983) Formal Spatial Analysis. Gower, Aldershot

Palander TF (1935) Beiträge zur Standorttheorie. Almquist and Wiksell, Uppsala

CHAPTER 8

# Localization

Localization refers to the concentration of an industry in one or a few locations.

## Resource Oriented Industry

This occurs first of all in resource oriented industries, which are localized to the extent that the resource deposits are. When there is but one resource deposit, the firms of a resource oriented industry will all be located there. In view of returns to scale – manifested by a fixed cost F – there may well emerge a single firm after an oligopolistic struggle e.g. a price war, or else an equilibrium exists in which a comfortable profit margin permits all contenders to recover their fixed cost.

There may also be a competitive market when firms encounter different marginal cost in the extraction of the resource due to variations in local conditions or quality of the resource. The price is then set by the marginal cost of the marginal firm. In view of the spatial concentration in one place, the non-spatial theory of firms and industry applies, about which nothing more need be said here.

With several resource deposits, each may be utilized by one or several firms competing under similar conditions as described above. With each resource deposit is then associated a market area served by the firms located at that deposit. Depending on the transportation cost of the product the market areas need not cover the entire region, however.

So far it was assumed that the resource deposit is itself concentrated in a point or small area. But this is not the case for agricultural land. The processing of agricultural raw materials typically requires a supply area for each firm. Sugar refineries are each located in the midst of fields of sugar cane or sugar beets, and the same applies to the canning or freezing of fruit and vegetables and in regard to rubber plantations, coconut groves etc.

A coal field may be similarly extended, but coal mines that go underground may be quite close together on the surface when mining different seams at different depths, while open pit mining clearly requires some spatial separation. In oil drilling there is often an uneconomical tendency for wells to be placed too close together in an effort to tap the same pool exploited already by a competitor's existing well.

Firms milling in transit will seek out the same nodal points and therefore tend to be localized there.

A labor oriented industry will go to areas of cheap labor, which may be more or less extended. Labor mobility often results in concentration in a small area, near airports or in so- called "development zones" endowed with customs privileges.

When the market for an industry's product is localized, then a market oriented industry will be also be localized. If this market is another industry then the two will locate together and form an agglomeration (chapter 9).

## Externalities

A more subtle cause of localization is external economies, first mentioned by Alfred Marshall (1890).

"When an industry has thus chosen a locality for itself, it is likely to stay there long: so great are the advantages which people following the same skilled trade get from near neighborhood to one another. The mysteries of the trade become no mysteries; but are as it were in the air, and children learn many of them unconsciously. Good work is rightly appreciated, inventions and improvements in machinery, in processes and the general organization of the business have their merits promptly discussed: if one man starts a new idea, it is taken up by others and combined with suggestions of their own; and thus it becomes the source of further new ideas. And presently subsidiary trades grow up in the neighborhood, supplying it with implements and materials, organizing its traffic, and in many ways conducing to the economy of its material.

Again, the economic use of expensive machinery can sometimes be attained in a very high degree in a district in which there is a large aggregate production of the same kind, even though no individual capital employed in the trade be very large. For subsidiary industries devoting themselves each to one small branch of the process of production, and working it for a great many of their neighbors, are able to keep in constant use machinery of the most highly specialized character, and to make it pay its expenses, though its original cost may have been high, and its rate of depreciation very rapid.

Again, in all but the earliest stages of economic development a localized industry gains a great advantage from the fact that it offers a constant market for skill. Employers are apt to resort to any place where they are likely to find a good choice of workers with the special skill which they require; while men seeking employment naturally go to places where there are many employers who need such skill as theirs and where therefore it is likely to find a good market. The owner of an isolated factory, even if he has access to a plentiful supply of general labor, is often put to great shifts for want of some special skilled labor; and a skilled workman, when thrown out of employment in it, has no easy refuge." (Marshall 1920, p. 271)

The external economies mentioned are of two types, residing in a specialized labor force and in facilities and activities that support the entire industry and involve indivisibilities or, for other reasons, show increasing returns to scale.

## Retail Stores

In retailing, localization occurs only for shopping goods, while outlets for convenience goods by definition are dispersed.

In medieval cities, city streets specialized in the various trades, also those which are now considered to produce convenience goods: bakers, butchers, tailors etc. A modern example is 47th Street in New York City, for diamond stores. Luxury goods were displayed by the various sellers in special halls, famous architectural landmarks, located in prominent places.

Religious articles are offered then and now, near shrines or pilgrimage centers, which incidentally also attract the begging industry.

## Gravitation to the Median: the Hotelling Effect

An unsuspected cause of localization was first brought to light by Hotelling (1929): Consider two mobile vendors of ice cream, selling on a beach of length L occupied by sun bathers at a uniform density. Prices are given and equal or quantity demanded is independent of price, but customers will buy from the nearest vendor. Where are the profit maximizing locations of the two entrepreneurs?

Given the location of seller 1, the other will locate next to her/him on the side of the larger market. Eventually both will move to the center and each in this way secure one half the entire market. The welfare optimum locations, which also divide the markets equally among the two sellers would be at one fourth and three fourths of the length L of the market. Hotelling and subsequently Smithies (1941) and others have pointed out that this gravitation to the median is observed also in other contexts such as product design and the platforms of political parties, causing an uneconomical or undesired uniformity.

The Hotelling effect does depend on the assumption of rigid demand and a limited size of the market (mathematical notes 8.2., 8.3).

The generalization from a linear to a two-dimensional market is straight forward only for Manhattan metric where the median is sought both in the north-south and east-west directions (say). For more than four firms no matter what the metric, new firms seeking maximum market areas will move away from the center. This process has been simulated many times and the evolving shapes of market areas have been studied (Eaton 1972, Eaton and Lipsey 1976). Under Euclidean distance, hexagonal tessellation's seem to emerge after a sufficient number of entries, when relocation of existing firms is also allowed. Of course market area maximization is the appropriate objective only when demand is inelastic.

### Mathematical Note 8.1
### Resource Oriented Industry

Let i denote both a firm and its location by the resource deposit.

| | |
|---|---|
| $F_i$ , $c_i$ | its costs |
| $m_i$ | its capacity |
| $a-q=p$ | demand function for the product |

Industry equilibrium under perfect (non-spatial) competition is then determined by social surplus maximization

$$\underset{x_i \geq 0}{\mathrm{Max}} \int_0^{\Sigma x_i} (a-q)^2 \, dq - \sum_i c_i x_i + \sum \mu_i (m_i - x_i)$$

where

$$q = \sum_i x_i \qquad 0 \leq x_i \leq m_i$$

This simple "capacitated" quadratic program is solved by

$$x_i \left\{\begin{matrix} = \\ \geq \end{matrix}\right\} 0 <=> \left(a - \sum_i x_i\right) \left\{\begin{matrix} < \\ = \end{matrix}\right\} c_i + \mu_i$$

or since

$$p = a - \sum_i x_i$$

is the market price

$$x_i \left\{\begin{matrix} = \\ \geq \end{matrix}\right\} 0 \Leftrightarrow p_i \left\{\begin{matrix} < \\ \geq \end{matrix}\right\} c_i$$

since the profit margin $\mu_i$

$$\mu_i \left\{\begin{matrix} = \\ \geq \end{matrix}\right\} 0 <=> x_i \left\{\begin{matrix} < \\ = \end{matrix}\right\} m_i$$

vanishes when $x_i = 0$.

Firms with marginal cost below price should produce at full capacity and those with marginal cost equal to price, a non-negative amount up to capacity. In the long run firms must also cover their fixed costs.

$$\mu_i h_i \geq F_i$$

or

$$(p_i - c_i) x_i - F_i \geq 0$$

for their outputs $x_i$. Price determination is described graphically in figure 8.1.

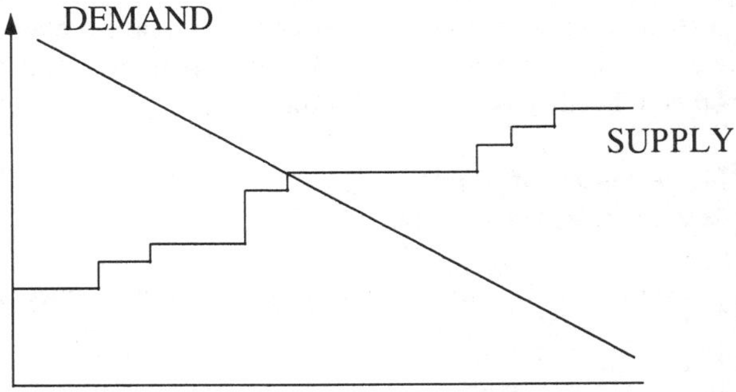


**Fig. 8.1.** Equilibrium in an industry with different costs and capacities of firms

### Mathematical Note 8.2
### Hotelling Model

A market oriented industry of just two firms is the subject of the famous location model of Hotelling (1929).

Imagine a beach of given length L occupied by sun-bathers at a uniform density. Let a first vendor of ice-cream be stationed at a point y. Where would a second vendor choose to locate when the demand per person is given and constant and buyers will always go to the nearest seller? (figure 8.1).

Let $y \geq \frac{1}{2}$. Then a second firm seeks a location $x \leq y$. The market is divided at $\frac{y+x}{2}$ and the second firm's market area is

$$x + \frac{y-x}{2}$$

Now

$$\max_{x \leq y} x + \frac{1}{2}(y-x) = \frac{y}{2} + \frac{1}{2}\max_{x \leq y} x = y$$

The answer is right next to y on the side of the wider line segment, for that captures the greatest market. y in turn will move to that side of x, the second firm's location, and so on, until both are straddling the midpoint, or the median in case of an uneven distribution of customers.

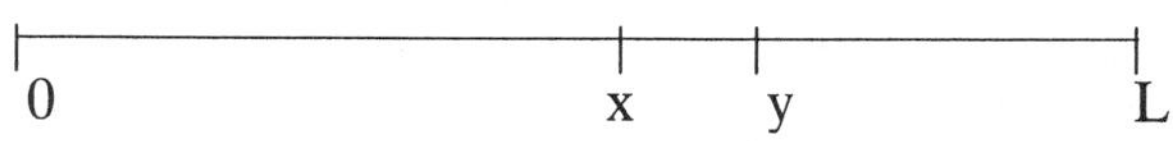


**Fig. 8.1.** Selling ice-cream on a beach

### Mathematical Note 8.3
### Hotelling Model for Linear Demand Functions

When quantity demanded depends on price, maximum and optimum market sizes exist for a firm. Therefore, if the regional market (L) is large enough, firms will keep a sufficient distance from each other in order to enjoy separation, that is non-overlapping optimal market areas. In the linear demand case for instance, suppose that

$$y > \frac{2}{3}\frac{a-p}{k} \tag{1}$$

at a mill price p given and equal for both firms. Consider the best location x for the second firm. At location x profits to the new entrant are

$$(p-c)\int_0^x f(p+kr)\rho dr + (p-c)\int_0^{\frac{y-x}{2}} f(p+kr)\rho dr \tag{2}$$

Maximizing profits with respect to x

$$f(p+kx) = \frac{1}{2} f\left(p + k\frac{y-x}{2}\right) \tag{3}$$

implying

$$p + k\frac{y-x}{2} < p + kx \tag{4}$$

or

$$x > \frac{y-x}{2}$$

$$x > \frac{2}{3}y.$$

For linear demand functions

$$a - p - kx = \frac{1}{2}\left(a - p - k\frac{y-x}{2}\right) \tag{3a}$$

$$x = \frac{2}{5}\left(\frac{a-p}{k} + y\right) < y \tag{5}$$

is located to the left of y, using (1).
But $x = y$
when

$$\frac{2}{5}\left(\frac{a-p}{k} + y\right) \geq y \tag{5a}$$

that is

$$y \leq \frac{2}{3}\frac{a-p}{k}$$

Thus joint location depends both on size of the potential market y and price. For any price dependent demand function it remains true that separation of market areas occurs in large markets and joint location in sufficiently small ones.

## References

d'Aspremont C, Gabszewicz JJ, Thisse J-F (1979) On Hotelling's Stability in Competition. Econometrica 47:1045–1050

David P, Rosenbloom JL (1990) Marshallian Factor Market Externalities and the Dynamics of Industrial Localization. Journal of Economics 28:349–370

Fujita M, Thisse J-F (1986) Spatial Competition with a Land Market: Hotelling and von Thünen Unified. Review of Economic Studies 53:819–841

Gabszewicz J, Thisse J (1986) Spatial Competition and the Location of Firms. In: Gabszewich, Thisse, Fujita (eds) Location Theory. Harwood: Academic Publishers

Hotelling H (1929) Stability in Competition. Economic Journal 39:41–57

Lerner A, Singer HW (1937) Some Notes on Duopoly and Spatial Competition. Journal of Political Economy 45:145–186

Marshall A (1920) Principles of Economics, 8th Edition. Mcmillan, London

Saxenian A (1994) Regional Advantage: Culture and Competition in Silicon Valley and Route 128. Harvard University Press, Cambridge, MA

Smithies A (1941) Optimal Location in Spatial Competition. Journal of Political Economy 49:423–439

Stahl K (1987) Urban Business Location. In: Mills E (ed) Handbook of Regional and Urban Economics

CHAPTER 9

# Agglomeration

This chapter concerns the spatial interaction of several industries. It can take two extreme forms: clustering, i.e. "agglomeration" or repulsion leading to disjoint locations and locational specialization (to be considered in chapters 10 and 11).

The term agglomeration was first coined by Alfred Weber (1905) to denote the mutual attraction of different industries resulting in their joint location.

In discussing agglomeration economies, some authors (Hoover 1936, ch.6, Isard 1956, ch.8, Fujita and Thisse 1995) have distinguished between "localization economies" for firms of the same industry in one location from "urbanization economies" of firms from various industries in one location. In this book we use agglomeration only in the sense of "urbanization economies".

The most obvious attraction is that between a supplier and the user of an intermediate good. This is the equivalent of market orientation when the market is another industry. A well-known example is the automobile industry attracting its various suppliers of tires, glass, electric equipment, upholstery etc. to adjacent locations. The Japanese "Kan ban" or "Just in time" system depends on the close proximity of suppliers to meet very stringent delivery schedules. A classical example is also that of tanneries near slaughter houses and ( more sobering) of cemeteries located next to hospitals.

In addition to savings in transportation cost, there may be energy economies, as in the case of a steel mill being placed next to a blast furnace, so that the molten iron can be transferred immediately to the converter of the steel mill. Vertical integration which often results is an extreme form of agglomeration.

When not related as supplier and users of products, firms from different industries can share the services of support facilities such as packaging, shipping, storage, repairs, training, marketing and finance, offered in one location. They have access to an industrial labor force and share the social overhead of a city (cf chapter 8 above).

A new industry may chose to locate next to an established one because there exists already a labor market and a social infrastructure. If the product is a consumer good, then the presence of other firms in a location means that a local consumer market already exists for these consumer products.

"A district which is dependent chiefly on one industry is liable to extreme depression, in case of a falling-off in the demand for its produce, or of a failure in the supply of the raw material which it uses. This evil again is in a great measure avoided by those large towns or large industrial districts in which several distinct industries are strongly developed. If one of them fails for a time, the others are likely to support it indirectly; and they enable local shopkeepers to continue their assistance to work-people in it." (Marshall, p.272).

Agglomeration economies in selling appear in the case of shopping goods. The attraction of different stores to each other, realized in shopping districts and shopping centers, results in part from the convenience of combining shopping trips for different but related goods. In part it is once more a matter of information, i.e. when shopping streets or shopping centers establish a reputation for offering goods in a particular quality class and thus a certain price range. At the high end one finds 5th Avenue in New York and Rodeo Drive in Los Angeles, at the lower end discount stores and factory outlets as for outer wear and knitted goods in Fall River, Ma.

"So far we have discussed localization from the point of view of the economy of production. But there is also the convenience of the customer to be considered. He will go to the nearest shop for a trifling purchase; but for an important purchase he will take the trouble of visiting any part of the town where he knows that there are specially good shops for his purpose. Consequently shops which deal in expensive and choice objects tend to congregate together; and those which supply ordinary domestic needs do not" (Marshall, p.273).

Clustering is observed also in entertainment activities. Thus in the US, ice-cream parlors and coffee shops are often attracted to movie theaters, and restaurants to the theater districts in New York, Boston and other cities.

Firms with market areas of different sizes, once located together in one place, will find themselves drawn together again in clusters elsewhere, as shown by the example of figure 9.1.

Consider a hypothetical case with market areas of radius 2, 3, 5, and 7 (say) in a linearly extended economy.

Let plants be spaced one unit apart in a linear region. Denoting them by their market radii of 2, 3, ... clusters arise through coincidence as seen from figure 9.1. If all plants are located together at one point to begin with, the clusters of various size will be repeated periodically. In fact, clusters emerge even when firms are at first separated. Lösch has shown how hexagonally structured nets of market areas may be rotated so that a maximum number of joint locations of industries is generated.

The Hakimi-Gülicher effect will also draw otherwise unrelated industries to important nodes in the transportation network, where they can then benefit from the economies of shared overhead facilities. Perhaps the most significant manifestation of agglomeration economies is the formation of cities, i.e. of concentrated human settlements, intended to share social overhead. Historically important was the sharing of defense. Protection is also at the core of police, fire fighting and health services, as well as sanitary facilities, including water supply.

Cities provide access to its citizens through road networks and public transportation. The social overhead of protection and accessibility operates under increasing

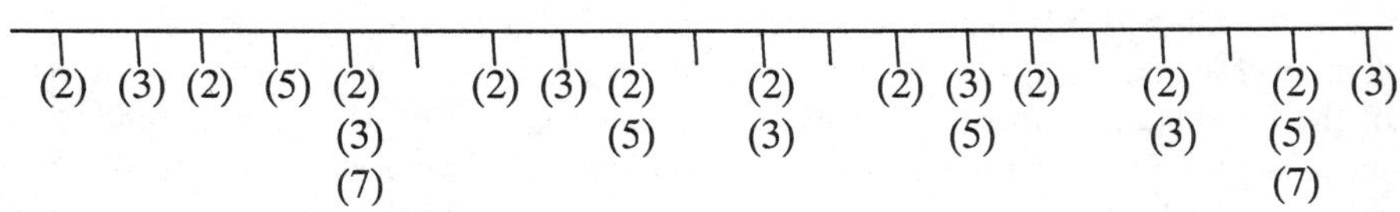


**Fig. 9.1.** Locations of industries with market radius of 2, 3, 5, 7.

returns to scale and is thus economically shared rather than produced individually for (or by) each user. Urban structure is considered in chapter 11 and urban systems in chapter 12.

### Mathematical Note 9.1
### Incentives for Joint Location

To illustrate agglomeration economies consider two firms whose markets are in locations $m_1$ and $m_2$ (figure 9.2). By locating together their unit production costs $c_1$ and $c_2$ can be reduced to

$$c_{12} < c_1$$

and

$$c_{21} < c_2$$

Sales are $q_1, q_2$ respectively

and unit transportation cost $k_1, k_2$

Fig. 9.1

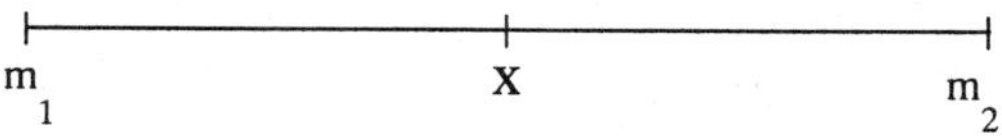


Let x be their joint location. The socially optimal choice is determined by

$$\underset{m_1 \leq x \leq m_2}{\text{Min}} \quad k_1 q_1 (x - m_1) + k_2 q_2 (m_2 - x)$$

$$= k_2 q_2 m_2 - k_1 q_1 m_1 + \underset{x}{\text{Min}} (k_1 q_1 - k_2 q_2) x$$

so that

$$x = \begin{Bmatrix} m_1 \\ m_2 \end{Bmatrix} \iff k_1 q_1 \begin{Bmatrix} > \\ < \end{Bmatrix} k_2 q_2$$

When $k_1 q_1 = k_2 q_2$ any location in $m_1 \leq x \leq m_2$ is equally good.

Suppose $x = m_1$. This is socially optimal provided total cost from joint location is less than that for individual locations $m_1$, $m_2$

or

$$(c_1 - c_{12}) q_1 + (c_2 - c_{21}) q_2 > (m_2 - m_1) k_2 q_2 \tag{1}$$

For firm 2 this is acceptable only if its savings on production costs exceed the newly incurred transportation costs (per unit)

$$(c_2 - c_{21}) > (m_2 - m_1) k_2 \tag{2}$$

a stronger requirement than (1). This illustrates that not all agglomeration economies may actually be realized by self-interested firms to whom they appear as external economies.

### Mathematical Note 9.2
### Intermediate Products

As a special case let cost savings from joint location arise from transportation costs for intermediate products. Suppose firm 1 uses $a_{21}$ units of product 2 as input per unit of its output and firm 2 $a_{12}$ units of product 1.

Let firms locate in x and firm 2 in y (figure 9.3)

Fig. 9.2. Location of two interacting firms

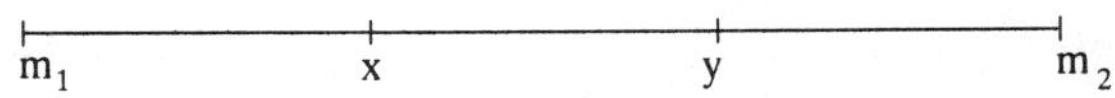


Total transportation cost is now

$$\begin{aligned} & k_1 q_1 (x - m_1) + k_2 a_{21} q_1 (y - x) \\ & + k_2 q_2 (m_2 - y) + k_1 a_{12} q_2 (y - x) \\ & = \text{const.} + [k_1 q_1 - k_2 a_{21} q_1 - k_1 a_{12} q_2] x \\ & + (k_1 a_{12} q_2 - k_2 q_2 + k_2 a_{21} q_1) y \end{aligned} \tag{1}$$

A joint location is optimal when the combined transportation costs of inputs exceed the smaller of the transportation costs for one or the other products. Then the joint location will be drawn to the market location of the product with the greater transportation cost. It is $m_1$ when

$$k_1 q_1 > k_2 q_2 \tag{2}$$

and

$$k_2 q_2 < k_1 a_{12} q_2 + k_2 a_{21} q_1 \tag{3}$$

Firm 2 will prefer location $m_1$ whenever transportation cost for its input is larger than for its output, i.e. when it is resource oriented

$$k_2 q_2 < k_1 a_{12} q_2 \tag{4}$$

which is a stronger requirement than (3).

Again this incentive may be too weak for firm 2 to locate together with firm 1.

Now let the supplier's (say firm 1's) market be the other firm (firm 2)

$$q_1 = a_{12}\, q_2$$

and the other firm's distribution costs in the product market dominate its transportation costs for inputs, i.e. firm 2 is market oriented.

$$k_2 > k_1 a_{12}$$

Potential competition from other firms in its industry will then force firm 1 to locate near firm 2. Firm 1 is then also market oriented. The best known example of this are the suppliers of parts to automobile assembly plants. It should apply also to the parts producers for other durable goods.

### Mathematical Note 9.3
### Agglomeration Economies Versus Scale Economies

The distinction between the two types is brought out clearly in the following case where firms may operate either a single plant or two plants in the presence of scale economies and of transportation costs for inputs.

Suppose the same firm can operate two types of plants in two locations, but available space permits only one plant of optimal capacity $x = 1$ in either location. This is the situation assumed in the assignment problems (cf mathematical note 10.1). Products are sold in competitive markets and are equally profitable in either location. Total production costs as a function of output are

$$c_i x_i^{\alpha} \quad 0 < \alpha < 1$$

so that unit costs

$$c_i x_i^{\alpha-1}$$

decrease with scale.

Let production of 2 require one unit of product 1 as input. No transportation cost k for the input is incurred when plants of capacity $x_i = \frac{1}{2}$ operate in either location. Comparing costs

$$2 \cdot \left(\frac{1}{2}\right)^{\alpha} (c_1 + c_2) \gtrless c_1 + c_2 + k$$

$$\left(2^{1-\alpha} - 1\right)(c_1 + c_2) \gtrless k$$

The left hand is savings from scale economies and the right hand savings from agglomeration economies. Either may be the larger one leading to separate locations and optimal sizes of plant or joint location at half the optimal size.

## References

Fujita M, Smith TE (1990) Additive-Interaction Models of Spatial Agglomeration. Journal of Regional Science 30:51–74

Fujita M, Thisse J-F (1995) Economics of Agglomeration, Eleventh World Congress of the International Economic Association, Tunis

Healy RG (1972) Agglomeration and Footlooseness: The Distribution of Economic Activities Among Metropolitan Areas. University of California Press, Los Angeles

Marshall A (1920) Principles of Economics, 8th edition. Macmillan, London

Papageorgiou YY, Smith TR (1983) Agglomeration as Local Instability of Spatially Uniform Steady-States. Econometrica 51:1109–1119

Papageorgiou YY, Thisse J-F (1985) Agglomeration as Spatial Interdependence Between Firms and Households. Journal of Economic Theory 37:19–31

Rivera-Batiz F (1988) Increasing Returns, Monopolistic Competition and Agglomeration Economies in Consumption and Production. Regional Science and Urban Economics 18:125–153

Scitovsky T (1954) Two Concepts of External Economies. Journal of Political Economy 62:143–151

Stahl K (1982) Differentiated Products, Consumer Search, and Locational Oligopoly. Journal of Industrial Economics 31: 97–113

Wang P (1993) Agglomeration in a Linear City with Heterogeneous Households. Regional Science and Urban Economics 23:291–306

Weber A (1909) Über den Standort der Industrien. J.C.B. Mohr, Tübingen. – Englisch translation: The Theory of the Location of Industries (1929) University of Chicaco Press, Chicago

Wolinsky A (1983) Retail Trade Concentration Due to Consumers' Imperfect Information. Bell Journal of Economics 14:275–282

# CHAPTER 10

# Assignment

Although agglomeration has been much discussed in the recent literature (Fujita and Smith 1990, Krugman 1991, Fujita and Thisse 1995,) we should not overlook the existence of forces working in the opposite direction, towards deconcentration or deglomeration. These forces result first of all from competition of industries for limited resources, specifically land. Locational specialization as a result of competition for land use is treated in chapter 11. Deglomeration may also result from spatial heterogeneity which itself is caused by unequal resource endowments or unspecified attributes determining the suitability of a location for particular activities. This is the subject of locational assignment to be explored in this chapter.

## The Linear Assignment Problem

To study the effect of heterogeneity, consider a number of plants from different industries and an equal number of locations to be chosen from, each location having a carrying capacity for one full sized plant. The suitability, i.e. profitability of the location i for plant k is assumed to be given: $b_{ik}$.

Since plants are assumed to operate under constant returns to scale, it is feasible also to distribute a plant over several locations and thus to bring plant operations together from different industries in any one location. Are agglomerations of this kind ever chosen or will full sized plants emerge occupying separate locations when aggregate profits are maximized?

This is the famous assignment problem, first formulated by the psychologist Thorndyke for assigning army personnel to army jobs and later considered for the assignment of plants to locations by Koopmans and Beckmann (1957).

It turns out that one to one assignments are superior in the sense that there is always one among the optimal solutions.

These assignments can be characterized by the rents that they yield to locations i , $\lambda_i$ and the quasi-rents $\mu_k$ they yield to plants k. Any combination ik in the optimal assignment will earn these rents for i and k, while this is not so for the non-optimal combinations

$$x_{ik} \begin{Bmatrix} = \\ \geq \end{Bmatrix} 0 \Leftrightarrow b_{ik} \begin{Bmatrix} \leq \\ = \end{Bmatrix} \lambda_i + \mu_k \tag{1}$$

From (1) it follows that

$$\lambda_i = \max_k b_{ik} - \mu_k \tag{2}$$

$$\mu_k = \max_i b_{ik} - \lambda_i \tag{3}$$

Equations (2) and (3) state the principle of comparative advantage which becomes an absolute advantage after allowing for rents: a firm maximizes its profits and earns quasi-rent $\mu_k$ at the optimally assigned location when location rent is deducted from gross profits $b_{ik}$ , and a landlord earns maximal rent by selecting the firm with the highest net profit after allowing for quasi-rent.

As an example consider the case of two plants and two locations and suppose we have found rents for an assignment 11, 22

$$b_{11} = \lambda_1 + \mu_1 \qquad b_{22} = \lambda_2 + \mu_2 \tag{4}$$

while

$$b_{12} < \lambda_1 + \mu_2 \qquad b_{21} < \lambda_2 + \mu_1 \;. \tag{5}$$

Then

$$\begin{aligned} b_{11} + b_{22} &= \lambda_1 + \mu_1 + \lambda_2 + \mu_2 \\ &= (\lambda_1 + \mu_2) + (\lambda_2 + \mu_1) \\ &> b_{12} + b_{21} \end{aligned} \tag{6}$$

proves that this was an optimal assignment.

If a plant k is more profitable than some other plant j in all locations

$$b_{ik} > b_{ij} \qquad \text{all } i$$

then at the optimal location for plant k

$$\begin{aligned} b_{ik} - b_{ij} &\le (\lambda_i + \mu_k) - (\lambda_i + \mu_j) \\ &= \mu_k - \mu_j \end{aligned}$$

Plant k has a higher quasi-rent than plant j. A similar statement can be proved for locations: If a location is more profitable than some other location for all plants, then this location earns a higher rent.

It should be noted from the efficiency conditions (4), (5) that a constant may be added to all $\lambda_i$ and subtracted from all $\mu_k$ without change in the equations/inequalities. Thus the rents and quasi-rents are not unique but determined only up to an additive constant.

Now let the number of locations m exceed the number of plants n. The constraints on locational occupancy must be relaxed to an inequality. It is clear that not all locations can be fully utilized any more and at those locations not used or underutilized rents must be zero. This removes the indeterminacy of rents and quasi-rents.

This is also true when the number of plants exceeds that of locations so that not all plants can be fully located and those not will earn quasi-rents zero.

## The Quadratic Assignment Problem

Let now plants k require inputs (in weight units) $a_{hk}$ from plants h and let the distance between locations i and j be $d_{ij}$ . With transportation cost proportional to weight times distance, what is the assignment of plants to locations that minimizes total transportation costs of interplant transactions? Consider once more an assignment $x_{ih}$ of plant h to location i and $x_{jk}$ of plant k to location j. Total transportation cost is then

$$\sum_{i,j,h,k} x_{ih} a_{hk} x_{jk} d_{ij}$$

a quadratic function in the $x_{ih}$.

This is to be minimized subject to constraints that all plants be located and all locations be utilized once more. When $x_{ih}$, $x_{jk}$ are restricted to integer value 0 and 1, this is the so-called Quadratic Assignment problem (Koopmans Beckmann 1957) which turns out to be one of non-polynomial complexity. Relaxing the integer constraints will now always generate fractional solutions, in fact

$$x_{ik} = \frac{1}{n}$$

So that under the optimal assignment all inputs are supplied locally.

This is of course, agglomeration once more. An equilibrium with plants separated among locations cannot be sustained by prices when plants interact as mutual suppliers (mathematical note 10.3).

## Location of Production in the Short Run

An allocation of production to locations, more involved than a straight assignment, is made in the short run, when plant capacities are already in place and the economic problem is one of their utilization. A study of this was the first application of linear programming to location theory (Beckmann-Marschak 1955). Subsequently Benjamin Stevens made important contributions to location theory by means of linear programming (1969).

The allocation of production activities to locations in the short run is best understood in the light of efficiency conditions of Koopmans applied to this multi-commodity capacity and demand constraint transportation problem.

A multi-facility firm or an industry of several plants seeking maximum profits under competitive conditions will set or discover prices for the different commodities involved in all locations. When considering a multi-facility single firm which might act as a monopolist, we assume as a counterforce rectangular demand functions in market locations of the type

$$q_i(p) = \begin{cases} m_i & \text{for } p \le a_i \\ 0 & \text{for } p > a_i \end{cases}$$

When considering a competitive industry the demand functions may be arbitrary

$$q_i(p) = f_i(p)$$

To stay within the linear programming framework the demand functions must be step functions

$$f_i(p) = m_{i,h} \quad \text{for} \quad a_{i,h-1} < p \leq a_{i,h}$$

with

$$m_{i,h} < m_{i,h-1}$$

For smooth demand curves the problem format shifts to nonlinear programming of a type encountered in mathematical note 4.3 on the equilibrium of trade centers.

Production of a commodity k takes place in locations i, where (efficiency prices) just cover cost. These costs are given constant unit cost plus any rents for reaching limited capacities. Rents as discounts may also have to be subtracted from demand prices when the quantity limit of demand at the reservations prices $a_{ih}$ are reached.

Transportation activities take place wherever the arbitrage condition is satisfied that price differences must cover transportation costs.

Production activities are in this way attracted to sites of cheap production costs or easy access to high priced markets. In this short run scenario, location and production decisions are identical.

### Mathematical Note 10.1
### Integer Assignments are Optimal

Theorem: An optimal solution to the linear program

$$\underset{x_{ik} \geq_v}{\text{Max}} \sum_{i,k} b_{ik} x_{ik}$$

subject to

$$\sum_i x_{ik} = 1 \quad k = 1, \ldots k \tag{1}$$

$$\sum_k x_{ik} = 1 \quad i = 1, \ldots k \tag{2}$$

can always be chosen to be integer valued

$$x_{ik} = \begin{cases} 0 \\ 1 \end{cases} \tag{3}$$

Let $x_{ik}$ be the amount of plant k assigned to location i. The theorem is true for n = 1, for then the equations (1), (2) imply $x_{11} = 1$. Assume the theorem to be true for n − 1. Now Dantzig's theorem of linear programming states that the number of active variables $x_{ik}>0$ can be chosen less than or equal to the number of effective constraints.

The 2 n constraints

$$\sum_{i=1}^{n} x_{ik} = 1 \qquad \sum_{k=1}^{n} x_{ik} = 1 \tag{(1), (2)}$$

contain one redundant one, since the statement

$$\sum_k x_{nk} = \sum_k \left(1 - \sum_{i=1}^{n-1} x_{ik}\right) = n - (n-1) = 1$$

is implied by the other 2 n-1 equation (1), (2). Hence we need at most 2 n-1 positive variables $x_{ik}$. In the matrix $(x_{ik})$ there is thus either a row or a column with only one positive element, which must be 1 by (1), (2). Let it be $x_{hm} = 1$. Dropping plant m and location h from the problem we have reduced it to dimension n-1 for which the theorem was true by induction hypothesis.

**Mathematical Note 10.2**
**Efficiency Conditions for the Linear Assignment Problem**

The Lagrangean of the assignment problem of 10.1 is

$$\sum_{ik} b_{ik} x_{ik} + \sum_i \lambda_i \left(1 - \sum_k x_{ik}\right)$$

$$+ \sum_k \mu_k \left(1 - \sum_i x_{ik}\right)$$

and the efficiency conditions are

$$x_{ik} \begin{Bmatrix} = \\ \geq \end{Bmatrix} 0 \quad <=> \quad b_{ik} \begin{Bmatrix} < \\ = \end{Bmatrix} \lambda_i + \mu_k$$

in terms of land rents $\lambda_i$ and firms' quasi-rents $\mu_k$.

From this

$$\lambda_i = \max_k b_{ik} - \mu_k$$

$$\mu_k = \max_i b_{ik} - \lambda_i$$

If constraints (10.1.1) and (10.1.2) are relaxed to

$$\sum_k x_{ik} \leq 1 \tag{1}$$

$$\sum_i x_{ik} \leq 1 \tag{2}$$

then

$$\lambda_i \begin{Bmatrix} = \\ \geq \end{Bmatrix} 0 \Leftrightarrow \sum_k x_{ik} \begin{Bmatrix} < \\ = \end{Bmatrix} 1 \tag{3}$$

$$\mu_K \begin{Bmatrix} = \\ \geq \end{Bmatrix} 0 \Leftrightarrow \sum_I X_{IK} \begin{Bmatrix} < \\ = \end{Bmatrix} 1 \tag{4}$$

The rents and quasi-rents are always non-negative if all $b_{ik} \geq 0$. They have the dimension of $b_{ik}$, that is, money.

Notice that for an optimal assignment $\hat{x}_{ik}$

$$\sum_{i,k} b_{ik}\hat{x}_{ik} = \sum_{i,k} (\lambda_i + \mu_k)\hat{x}_{ik}$$

$$= \sum_i \lambda_i + \sum \mu_k \tag{5}$$

since each row and each column of the matrix $(\hat{x}_{ik})$ contains just one element 1 and only zeros otherwise.

The rents and quasi-rents add up to the full value of the optimal assignment. Hence their total is at a maximum. Searching for optimal assignments may thus also be called "rent seeking".

### Mathematical Note 10.3
### The Quadratic Assignment Problem: An Example

Let $a_{12} = a_{23} = a_{31} = 1$
and $a_{hm} = 0$ otherwise
Assume an assignment as in figure 10.3.

**Fig. 10.3.** A tentative assignment

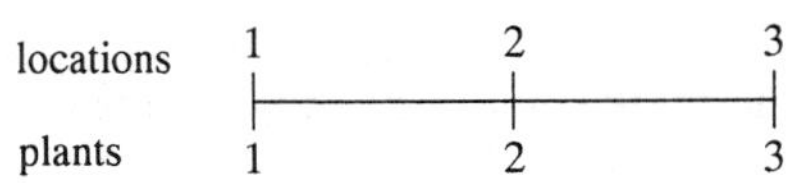


Transportation costs of inputs are then

| plant | 1 | 2 | 3 |
|---|---|---|---|
| costs | 2 | 1 | 1 |

In locational equilibrium rents should then be

| location | 1 | 2 | 3 |
|---|---|---|---|
| rent | 0 | 1 | 1 |

Now plant 3 would prefer location 1 at which its transportation costs are the same but rent is less. When plant 1 is forced out and moved to location 3 it would demand land rent to be lowered to zero, while land lord 1 would raise rent to 1. The game is now repeated with plant 3, preferring location 3 again. Only plant 2 is satisfied with its location, while plants 1 and 3 can never find a location to their satisfaction, i.e. are never in a state of equilibrium.

## Mathematical Note 10.4
## Location of Production in the Short Run: Plant Utilization

This note presents only the simplest version of location in the short run: a single product and given rectangular demand functions. Consider a firm whose plants i in locations i with capacity $c_i$ produce the same product to be sold in markets j. At given prices $p_j$ market demand in j is $q_j$ . Unit production costs in plant i are $h_i$ and transportation costs are $k_{ij}$. How should production be allocated to the plants in order to maximize profits? Since plants are located already and have given capacities, this must be considered as a location of production in the short run.

With the technology and market conditions as specified this is a linear program, a slight generalization of the well-known transportation problem (Hitchcock 1942, Koopmans 1949).

Denoting outputs by $x_i$ and shipments by $x_{ij}$

$$x_i = \sum_j x_{ij} \tag{1}$$

the profit maximizing choice is to

$$\underset{x_{ij} \geq 0}{\text{Max}} \sum_{ij} p_i x_{ij} - \sum_{ij} c_i x_{ij} - \sum_{ij} k_{ij} x_{ij} \tag{2}$$

$$\sum_j x_{ij} \leq c_i \qquad \text{for all i} \tag{3}$$

$$\sum_i x_{ij} \leq q_j \qquad \text{for all j} \tag{4}$$

With a Lagrangean

$$L = \sum_{ij} \left[ p_j - h_i - k_{ij} \right] x_{ij} + \sum_{ij} \lambda_j \left( q_j - \sum_i x_{ij} \right) + \sum_i \mu_i \left( c_i - \sum_j x_{ij} \right) \tag{5}$$

the Koopmans efficiency conditions are

$$x_{ij} \begin{Bmatrix} = \\ \geq \end{Bmatrix} 0 <=> p_j - \lambda_j \begin{Bmatrix} < \\ = \end{Bmatrix} c_i + \mu_i + k_{ij} \tag{6}$$

$$\lambda_j \begin{Bmatrix} = \\ \geq \end{Bmatrix} 0 <=> \sum_i x_{ij} \begin{Bmatrix} < \\ = \end{Bmatrix} q_j \tag{7}$$

$$\mu_i \begin{Bmatrix} = \\ \geq \end{Bmatrix} 0 <=> \sum_j x_{ij} \begin{Bmatrix} < \\ = \end{Bmatrix} c_i \tag{8}$$

According to (6) market j is supplied from plant i if and only if a non-negative profit margin $\lambda_j$ can be earned there after covering production and transportation cost plus (capacity) rent $\mu_i$ of the plant i. In fact to earn this profit $\lambda_j$, marketing managers seek for each j the cheapest source of supply, since (6) implies

$$\lambda_j = p_j - \operatorname*{Min}_i c_i + \mu_i + k_{ij} \tag{6a}$$

A market j is not served at all when $\lambda_j = 0$ and

$$p_j < \operatorname*{Min}_i c_i + \mu_i + k_{ij} \tag{6b}$$

Looked at from the plant managers' side, the plant rent (or plant profit) $\mu_i$ is earned by selecting the most lucrative markets

$$\mu_i = -c_i + \operatorname*{Max}_j \left(p_j - \lambda_j - k_{ij}\right) \tag{6c}$$

and plant i is laid idle whenever its production cost is too high

$$\mu_i = 0$$

and

$$h_i > \operatorname*{Max}_j \left(p_j - \lambda_j - k_{ij}\right) \tag{6d}$$

Rents $\mu_i$ or profit margins $\lambda_j$ are earned only when the plant is used to capacity and the sales quota $q_j$ is filled, respectively.

## References


Beckmann MJ, Marschak T (1955) An Activity Analysis Approach to Location Theory. Kyklos 8: 125–141. Also Second Symposium Linear Programming I (1955). National Bureau of Standards, Washington, DC, p 331–380

Dantzig G (1963) Linear Programming and Extensions. Princeton University Press, Princeton

Florence PS (1939) Report on the Location of Industry. PEP, London

Hartwick JM (1974) Price Sustainability of Location Assignments. Journal of Urban Economics 1:147–160

Koopmans TC, Beckmann M (1957) Assignment Problems and the Location of Economic Activities. Econometrica 25:53–76

Krugman PR (1993) Geography and Trade. Cambridge, Ma.: MIT Press

Lösch R (1994) The Economics of Location. Yale University Press, New Haven, p 369

Stevens B (1969) Linear Programming and Location Rent. Journal of Regional Science 3, 2:215–216

# CHAPTER 11

# Land Use

In land use theory the original question of location theory what are the locations of an economic activity is reversed: what are the economic activities at a location? This is the preferred approach for land intensive activities such as agriculture and housing, giving rise to agricultural and urban land use theories.

## Agricultural Land Use

The father of agricultural land use theory (as of marginal productivity theory) is Johann Heinrich von Thünen, who wrote in 1829 as follows

> Consider a very large town in the center of a fertile plain which does not contain any navigable rivers or canals. The soil of the plain is assumed to be of uniform fertility which allows cultivation everywhere. At a great distance the plain ends in an uncultivated wilderness, by which this state is absolutely cut off from the rest of the world.
>
> This plain is assumed to contain no other cities but the central town and in this all manufacturing products must be produced; the city depends entirely on the surrounding country for its supply of agricultural products.
>
> All mines and mineral deposits are assumed to be located right next to the central town.
>
> The question now is: How under these circumstances will agriculture be developed and how will the distance from the city affect agricultural methods when these are chosen in the optimal manner?[1]

We begin with a simpler question: how far will agriculture, considered as a single undifferentiated economic activity, be extended, if labor and land can be used in any combination to produce an output according to a standard production function with constant returns to scale?

Let the constant product price p in the city market be given as well as the wage rate w, the same in all locations, and let there be a constant rate of transportation cost k per unit commodity and unit distance. The net revenue per commodity unit produced at distance r is then a linear decreasing function of distance p-kr. The farmer as

[1] Johann Heinrich von Thünen, *Der isolierte Staat in Beziehung auf Nationalökonomie und Landwirtschaft* (Stuttgart: Gustav Fischer, 1966) pp. 11–12. (This is a reprint of the 1826 edition) – my translation.

land owner considers his return per unit of land achieved by an optimal intensity of production through a profit maximizing use of labor. The relevant variable is now the amount of labor employed per unit of land, the intensity of land use. The profit maximizing intensity is that at which the value of marginal product just equals the constant wage rate. A positive amount of labor is applied to land as long as the price received after transportation cost is itself positive.

The intensity of land use falls with distance and the limit of the cultivated area is determined as precisely that distance at which transportation costs incurred exhausts the price received in the market.

When labor and land must be used in fixed proportions, the limit of cultivation is instead that distance at which labor cost plus transportation cost per unit commodity just equals the price received in the market.

Consider also the possibility that at any distance the rate of transportation cost is itself proportional to the price of the agricultural product at that distance, because transportation cost consists of feed for draft animals and food for workers. Then the net revenue per commodity unit after transportation cost, i.e. the local market price, falls at a rate proportional to itself and thus becomes a negative exponential function of distance which never reaches zero – thereby permitting an unlimited expansion of agricultural land use. However, this is a rather exceptional case.

## von Thünen's Rings

von Thünen analyses what methods of agricultural production or land use should be employed at various distances from the market. This may now be considered a forerunner of parametric linear programming (Dantzig 1963). Since only one constraint is involved, that on land availability at any distance, the analysis can be carried out by means of a diagram (figure 11.2 p. 131).

Consider just two competing products, cabbage and wheat, and assume fixed coefficients of production. Prices received net of labor and transportation costs are then described by two straight lines, whose downward slopes represent the transportation cost per unit distance of the output of one land unit, say one acre. Here the cabbage grown on one acre has greater weight, hence higher transportation cost and thus a steeper slope. At any particular distance the higher line indicates the more profitable product. It is possible that one line is always beneath another, as in the case of flax compared to all others (figure 11.2). Then this product will nowhere be grown for profit. Where the two lines intersect, the steeper is higher to the left and the flatter is higher to the right of the intersection. The intersection shows the precise distance at which profit oriented farmers will switch from the heavier to the lighter commodity (in terms of acre product). As this comparison is extended from two to more commodities, a sequence of zones is discovered, each zone being a circular ring, in which land is used for one particular product or method of cultivation exclusively. A typical sequence is vegetables, potatoes, corn (maize) wheat, pasture, forest. Notice that although wood, the forest product, has higher specific weight, its annual growth per acre incurs less transportation cost than vegetables.

The upper contour of the net revenue lines represents maximum net revenue or rent per acre. The rent is thus seen to decrease with distance from the city at a rate which is constant in each ring and decreases from ring to ring. This contour is therefore "convex". Eventually, at the outer border of cultivation, rent falls to zero.

Advances in transportation technology since von Thünen's time have made it economical to ship fresh vegetables grown in semi-tropical Florida to markets in the industrial North, across the other land use zones. The inner ring is now more often one of dairy farms, producing fresh milk for urban markets.

The von Thünen model explains a number of important facts about agricultural locations – in a strikingly simple way; particularly the fact of agricultural specialization even in the absence of climatic and fertility differences, and the decrease of rent with distance from markets. It is a fertile model capable of extension and modification in various ways. Some of these will now be indicated.

## Quantitative Demand

To determine the price level and the extent of cultivation of this "isolated state", consider the size of the urban population and the quantities demanded of various food stuffs and other agricultural products (fibers) as given. Then prices must be found such that the required quantities are forthcoming. The order of cultivation is again given by the ranking of weights produced per acre. The quantities of products can then be translated into acreages and the distance found where the various rings begin. However, care must be taken that the maximal distance required is technologically attainable, that is, that the inputs for transportation do not equal or exceed the resources of the isolated state. Obviously, at any given state of technology and resource availability, there exists a maximal size of the supply area of the city, and hence, a limit to the size of the city itself. Before this maximum is reached, however, other forces will usually cause a multiplication of cities and a decrease in the size of supply areas (see chapter 12).

If the isolated state is hemmed in by the "surrounding wilderness" and the population grows to such an extent that the available land area is taken up by the more intensive production activities, then agricultural products with the largest land input coefficient (e.g. wood or meat from grazing cattle) otherwise produced in an outer zone, will have to be imported.

A labor intensive product that was unprofitable before may now surface in a new ring. But the fact of rings with specialized land use remains.

## Variable Proportions

A more interesting variation concerns the nature of agricultural technology. (von Thünen assumed fixed input coefficients before Walras and Leontief). But is this essential? What if we assume substitutability between land and labor (and capital as well) as described by neo-classical production functions?

It then becomes a question of returns to scale. With constant or increasing returns to scale in local production, land use remains specialized and only a function of prices received and the wage rate. The order of cultivation becomes that of falling rates of transportation cost for the product produced per unit of land by one unit of labor (mathematical note 11.3).

With decreasing returns to scale it is economical to carry out several production activities simultaneously (i.e. in closest proximity) on one unit of land.

## Local Consumption

Should households located in the country purchase their agricultural products from suppliers in interior rings? No, for it is cheaper to produce these locally for local consumption, while it is still cheaper to buy products from outer rings as seen by comparing the rent that is yielded locally by products from other zones.

What about industrial products, those produced in the central city? We may assume that these can be transported free to rural customers as back freight on vehicles used to transport agricultural products to the city. Hence they should be available at a uniform delivered price to customers throughout the agricultural supply area, which is the market price in the metropolis.

Beyond the limits of the agricultural supply areas for the central city, subsistence agriculture is still economical, that is the growing of agricultural products for local consumption only. Land can then be cultivated even outside the supply areas of a city. But these subsistence farmers, unable to reach the market with their products, must do without industrial products from the city.

## Soil and Climate

What is the effect of quality differences in soil or climate? Given constant returns to scale in agriculture these will cause in turn a specialization of land use and thereby either eliminate or distort the von Thünen sings, or give rise to entirely new zones of land use. An example might be a coal field or oil field exploited right in the midst of surrounding agricultural zones. A particular spatial pattern cannot be predicted. The principle still holds, however, that the rent bid for the most productive (profitable) use will determine the use of land (see below).

The type of agricultural import depends on comparative advantage – it need not be that of an outer zone. However, if land is scarce it is likely that imports will consist of the more land-using products, that is, those produced in the more distant rings.

## Wage Differences

Wages may vary with distances from the city. This will affect labor costs of growing the various crops, and it will cause the slopes of the net revenue curves to differ from the transportation costs. Thus, if the wage level falls with increasing distance from the city, the net revenues will be tilted in an upward direction, causing rents to fall less rapidly than transportation costs of area products, and the supply areas to be larger.

## Transportation Facilities

Waterways, roads, or railroads will make transportation cheaper along some routes. The curves of equal distance or transportation costs ("isotims") are then no longer circles. In the case of three roads converging on the city and on the assumption that transportation costs on the road are half of those outside, the isotims are triangles whose vertexes lie on the trunk roads. If the transportation cost ratio is less than one-half, the isotims are star-shaped (Figure 11.2); if not, they approach hexagons.[2]

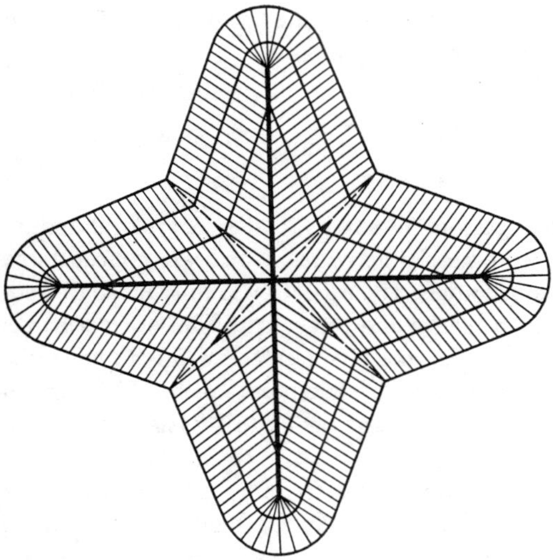

**Fig. 11.1.** Isotims and flows at an intersection of roads (from Beckmann 1968)

## The Isolated State as a Spatial Equilibrium System

The "Isolated State" of von Thünen is a spatial system capable of equilibrium and growth. To close this system we must include industrial production and transportation as economic activities explicitly. It is convenient to consider agriculture as producing a single product locally from labor and land according to a Cobb Douglas production function. One can then construct a macro-economic production function for the agricultural sector yielding aggregate output as a function of its aggregate labor input.

Capital as a factor of production is ignored. Thus transportation is assumed to be produced only from labor in fixed proportions. A single industrial good is produced in the city with labor according to a Cobb Douglas production function. Were capital included, this production function would have increasing returns to scale. With respect to labor alone, returns are decreasing.

Households are assumed to have identical tastes described by a log linear utility function. Welfare maximization achieved under perfect competition implies now that

[2] On isotims, cf. Tord Palander, Beiträge zur Standorttheorie (Uppsala: 1935).

individual and aggregate incomes are spent on the agricultural and the industrial product in fixed proportions. It can then be calculated that the entire labor force is allocated to agriculture, transportation and industry in proportions that are independent of the aggregate labor supply and depend only on the exponents of the production functions and the coefficients of the utility function. As population grows, the agricultural area expands but at a slower rate since the intensity of agricultural land use (labor employed per land) also increases. Any movement of labor from agricultural into industry must then be explained as a result of technical change.

The growth process is slowed down and the relations of the sector cease to be simple once the expansion of agricultural land is stopped by the "wilderness" that hems in the isolated state (mountains or deserts as definite borders).

## Urban Land Use

The residential areas of a city, comprising about 90% of its total area, are oriented towards the locations of industrial and commercial activities, which are concentrated through agglomeration economies in small areas. Their location is influenced by geographical features, the presence of a river or lake and the constitution of river banks and lake sides. Where no special features exist, the business area tends to be in the heart of a city. In larger urban areas a system of business centers in which production and distribution activities are clustered, emerges, analogous to a system of central places (to be discussed in chapter 12).

It is distance to the (nearest) business center that determines the rental value of housing and the space demanded by households, its inverse being the residential density of households. In turn the rental value of housing determines land rent, and the amount of housing space constructed on a unit of land.

To determine the demand for housing space requires an analysis of household preferences and consumer decisions. This analysis is usually cast in terms of the maximization of a representative household's utility function. The arguments of this utility functions should include housing space and general consumption in any case. One may also consider leisure, i.e. time left after daily work and commuting. In the simplest case a household's choice of location involve only housing space and general consumption (equal to income minus housing rent minus expenditure for commuting and shopping trips). Both commuting and shopping are directed to the nearest business center. The analysis is simplest if there is but one such center, as is true for small and medium sized cities in the United States, and the city is then said to be "monocentric". Locational choice reduces to the choice of a distance from the center, assuming that no preference exist with respect to directions (the isotropic case). As distance is increased, commuting expenditure rises but rents must fall to compensate for this and cause every location (distance) to be demanded by some household. Again to simplify the analysis assume initially all households to have the same income and tastes, and ignore all attributes of housing other than housing space – or alternatively let desirable attributes be increasingly available with space. (Each branch of economics has its own fictions, that is assumptions that are considered defensible. Equal incomes is one in urban economics).

A household may thus trade off housing space for general consumption in choosing its distance, with the result that more space is occupied by a household the farther it chooses to live. The lesser consumption of other goods is the result of rent for more space and of increased expenditure on transportation in commuting. If the expenditure on housing tends to be a fixed proportion of a household's income, about 25% in the US, then it is the trading of space for general consumption through increased transportation cost..

When the preferred amount of space is rented at a chosen distance, then on the assumption of equal tastes and incomes, the achieved utility must be the same in all locations for all households: households must be indifferent among locations. If time is also an argument in the utility function, the choice of location, i.e. of distance, involves a trade off of housing space for both time (of commuting) and general consumption. Nothing essential is changed, however.

One may next consider two income classes bidding for housing space. Richer households are less sensitive to the money cost of commuting: they outbid the poorer at longer distances and are outbid at shorter distances so that poorer households can live nearer to the business center in more crowded space but with shorter distances to work. The richer household enjoy more space but at the cost of larger commuting distances.

Household composition also plays a part in residential choices. Households with two or more commuters are more sensitive to distance than a one-commuter household and will outbid them at shorter distances. The structure of urban areas in contemporary America is explained by these simple facts: suburbs as the preferred location of middle income families with children living in ampler housing space while the inner city is sought by both poorer families and by yuppie households of two commuters without children.

The analysis goes beyond this to determine the levels of rent and household density as functions of distance for a monocentric city from specific utility functions to yield specific residential densities.

It has been observed, first by Colin Clark, that residential density falls off exponentially with distance from the center in many cities worldwide (Colin Clark 1945). This can be explained as follows. Assume that utility is an additive logarithmic function with respect to space and general consumption. Then it follows first that expenditure for housing is a constant proportion of income. Then, an approximate exponential law follows for density as a function of distance (mathematical note 11.5).

Is there any connection between von Thünen's theory of agricultural land use and the urban land use theory for the monocentric city?

In both cases economic activities and hence land use, are oriented towards a center, the metropolis or the CBD, respectively. The flows considered are of agricultural products in one case and of commuters in the other. But what of the von Thünen's rings? Is there an analog in urban housing? What is needed then are fixed coefficients for commuters housed per area of land when different housing types are considered.

To the weight of the product grown on one acre corresponds now the number of commuters dispatched from housing located on one acre of ground space. With each type of residential building is associated one such output (or inverse input) coefficient just as there was one land input coefficient for each type of agricultural product. von Thünen's theory predicts that the city should be composed of rings, in each of which there is just one type of residential building, and that their sequence should be in the order of the commuter numbers that can be housed on one unit of ground space. This implies specifically others that the height of buildings should decrease from ring to ring, as one moves outward. The choices of landlords are guided by commuters' willingness to pay for access to the center exactly as farmers are guided by net commodity prices received. In this approach the preference of households for space and general consumption or leisure are ignored, however, and a planning view of "housing needs" (as so many square meters per household) is taken tacitly. Of course, housing types should be distinguished in the housing supply function and the monocentric city model with consumers' choices should be formulated with supply as well as demand.

In particular, housing rent and land rent must be distinguished. To convert housing rent into land rent one must introduce profit maximizing land lords, who choose a supply of housing space per land unit by constructing a residential building that maximizes the surplus of housing rent over the (pro-rated) interest and maintenance cost of the building. This surplus becomes their rent bid for land and is the (maximal) income to the owner of the land. The housing rent level, which is at its maximum next to the central business district, is determined finally by the aggregate demand for housing of a city population given either exogenously or determined by the labor demand of local industries. The greater the urban population, the higher the housing and land rent level in the city.

## Bidding for Land Use

Land use can be analyzed without reference to a central city as market location. Consider any spatially extended market in which prices exist at all locations. Local prices for commodities as inputs and/or outputs and local wage rates allow an entrepreneur to calculate the profits that might be achieved by an economic activity and the amount of land needed in a certain location. Profits per unit of land are then the maximum rent that this enterprise could pay, and this defines a rent which that enterprise can bid for at this location. Competition for land use allocates land to the highest bidder, so that land use results in the maximization of land rent.

The bid rent is generally composed of two parts, one defined by the special qualities of land, its fertility for agricultural use or its agreeable features for residential use (natural beauty, view, climatic features) and an accessibility component, its proximity to relevant sources of supply and/or demand. Allocation to the highest bidder results in specialization of land use. The von Thünen rings generalize into extended areas of specialization by types of production or types of consumption (residential) uses.

### Mathematical Note 11.1
### Extent of Cultivation

Let $L$ = labor
$M$ = land
$Q = F(x, y)$ agricultural product of a single homogeneous good

with F a linear homogeneous production function.

The price received after transportation cost per product unit is

$$p - kr$$

and profits per unit of land are a function of the labor land ratio

$$\frac{1}{M}\left[(p-kr)F(L,M) - wL\right]$$

$$= (p-kr)F\left(\frac{L}{M},1\right) - w\frac{L}{M}$$

$$= (p-kr)f(\ell) - w\ell \qquad \frac{L}{M} = \ell \qquad \text{(say)}$$

Maximizing profits by choosing the labor land ratio $\hat{\ell}$ yields

$$0 = (p-kr)f'(\hat{\ell}) - w$$

$$f'(\hat{\ell}) = \frac{w}{p-kr} \tag{1}$$

Differentiate with respect to r

$$f''\frac{d\hat{\ell}}{dr} = \frac{wk}{(p-kr)^2}$$

$$\frac{d\hat{\ell}}{dr} = \frac{wk}{(p-kr)^2 f''} < 0 \tag{2}$$

since $f'' < 0$ by diminishing returns to substitution

The intensity of land use falls with distance and reaches zero where

$$f'(\hat{\ell}) = \infty \quad \text{at} \quad r = R = \frac{p}{k} \tag{3}$$

In the case of Cobb Douglas production functions

$$F = L^{\alpha}M^{\beta} \qquad \alpha + \beta = 1$$

$$f(\ell) = \ell^{\alpha}$$

$$(p-kr)\alpha\,\ell^{\alpha-1} = w$$

so that

$$\ell = \left[\frac{p-kr}{w}\right]^{\frac{1}{\beta}} \tag{4}$$

and output

$$q = \left(\frac{p - kr}{w}\right)^{\frac{\alpha}{\beta}} \tag{5}$$

Both labor input and product output increase with the price/wage ratio and fall with distance.

The limit of cultivation is at

$$R = \frac{p}{k} \tag{6}$$

Let now transportation cost be proportional to local agricultural price

$$k = k(r) = h \cdot p(r)$$

Then price received is

$$p\text{-}kr = p(r) \qquad \text{(say)}$$

$$p(r) = p - h\int_0^r p(t)dt \tag{7}$$

where p as before is the price in the city market.

Differentiating (7) with respect to r

$$p'(r) = -hp(r)$$

solved by

$$p(r) = pe^{-hr} \tag{8}$$

The price received for the agricultural product never falls to zero. The agricultural supply area of the city is unlimited.

Consider also production with fixed coefficients. Let a units of labor per land area be required to produce one unit of product per land area. With linear transportation costs profits are

$$p - a\,w - kr$$

Now the limit of the supply area is at distance

$$R = \frac{p - aw}{k} \tag{9}$$

Notice the difference between (6) and (9).

### Mathematical Note 11.2
### The von Thünen Model as a Linear Program

Consider

| | |
|---|---|
| $j = 1, \ldots n$ | products with |
| $a_j$ | input coefficients for land, and |
| $w_j$ | labor costs per unit output |

At prices $p_j$ in the central market, and
transportation cost $k_j$ per product unit and per unit distance

we seek to determine
the product output $x_j \geq 0$ of a farm of b acres located at distance r.

The farmers profits r are

$$\sum_j \left[p_j - w_j - k_j r\right] x_j \tag{1}$$

to be maximized under the constraint on land use

$$\sum_j x_j a_j \leq b \tag{2}$$

The Lagrange function is

$$\sum_j \left[p_j - w_j - k_j r\right] x_j + \lambda \left[b - \sum_j a_j x_j\right] \tag{3}$$

Its maximization yields the efficiency conditions

$$x_j \begin{Bmatrix} = \\ \geq \end{Bmatrix} 0 <=> p_j - w_j - k_j r \begin{Bmatrix} < \\ = \end{Bmatrix} \lambda a_j \tag{4}$$

The activity levels $x_j$ and rent $\lambda$ are functions of r. From (4)

Rent $$\lambda = \lambda(r) = \max_j \frac{p_j - w_j - k_j r}{a_j} \tag{5}$$

is a piece-wise linear decreasing and convex function of distance from the market r.

At every distance, the product with maximal yield per land unit is produced exclusively

$$j = \arg \max_i \frac{p_i - w_i - k_i r_i}{a_i}$$

As r increases, transportability, i.e. a low $k_j$ becomes ever more important. The diagram (figure 11.3) shows that the zones of cultivation follow in the order of decreasing transportation cost per product on one land unit $k_j / a_j$. As assumed in figure 11.2, cabbages weigh more than the wheat that can be grown on the same acre per year.

**Fig. 11.2.** von Thünen model for three activities

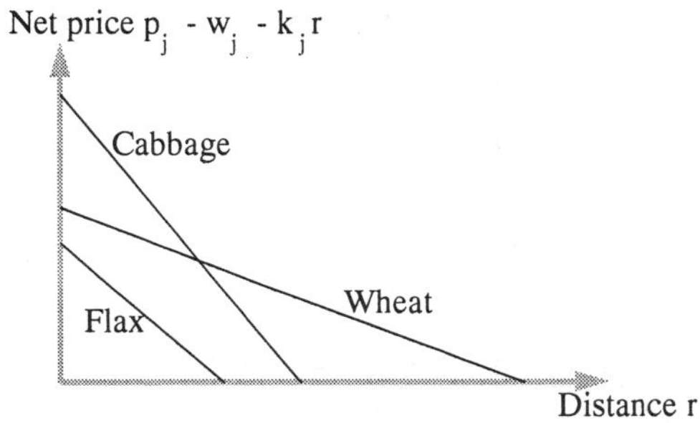

**Mathematical Note 11.3**
**Neo-classical Production Functions**

A constant returns Cobb Douglas function

$$Q_j = b_j L^{\alpha} M^{\beta} \qquad \alpha + \beta = 1 \tag{1}$$

for output $Q_j$ land input M and labor input L is reduced to

$$q = b\ell^{\alpha} \tag{2}$$

in terms of intensities of output per land

$$\frac{Q}{M} = q \tag{3}$$

and labor employed per land

$$\frac{L}{M} = \ell \tag{4}$$

The farmer must now choose which products to grow that is, what proportion $x_j$ of land to devote to product j subject to a constraint

$$\sum_j x_j \le 1 \tag{5}$$

and what labor intensity $\ell_j$ to apply. The profit per area to be maximized is

$$\max_{x_j \ge 0, \ell_j} \sum_j \left[ b_j \ell_j^{\alpha} - w_j \ell_j \right] x_j \tag{6}$$

subject to

$$\sum_j x_j \le 1 \tag{5}$$

maximizing the Lagrange function

$$L = \sum_j x_j b_j \left[ p_j - k_j r \right] \ell_j^{\alpha} - x_j \ell_j w \tag{7}$$

$$+ \lambda \left[ 1 - \sum_j x_j \right]$$

yields

$$x_j \begin{Bmatrix} = \\ \ge \end{Bmatrix} 0 \iff b_j \left[ p_j - k_j r \right] \ell_j^{\alpha} - w\ell_j \begin{Bmatrix} < \\ = \end{Bmatrix} \lambda \tag{8}$$

The optimal labor intensity is found from

$$0 = \frac{\partial L}{\partial \ell_j} = \alpha\, x_j b_j \left[p_j - k_j r\right] \ell_j^{\alpha-1} - x_j w \tag{9}$$

For products actually grown: $x_j > 0_j$ labor intensity is determined as

$$\alpha b_j \left[p_j - k_j r\right] \ell_j^{-\beta} = w$$

or

$$\ell_j = \left( \frac{\alpha b_j \left[p_j - k_j r\right]}{w} \right)^{\frac{1}{\beta}} \tag{10}$$

and profits per land unit from product j are found by substituting (10) in (6)

$$g = \beta b_j^{\frac{1}{\beta}} \left[p_j - k_j r\right]^{\frac{1}{\beta}} \left( \frac{\alpha}{w} \right)^{\frac{\alpha}{\beta}} \tag{11}$$

Both are clearly decreasing with distance r.

Rent $\lambda$ and the profit maximizing product j are determined by

$$\lambda = \max_j g_j = \left( \frac{\alpha}{w} \right)^{\frac{\alpha}{\beta}} \cdot \max_j \beta \left( b_j \left[p_j - k_j r\right] \right)^{\frac{1}{\beta}} \tag{12}$$

The criterion of choice is therefore

$$b_j \left(p_j - k_j r\right) \tag{13}$$

This criterion represents net prices received per unit product

$$p_j - k_j r \tag{14}$$

times land productivity $b_j$ , which is the output resulting from one unit of labor on one unit of land. $b_j k_j$ is the rate of transportation cost of this output of one labor unit on one unit of land.

Observe that (13) or (14) are not the same net price that appeared in the original von Thünen problem.

$$p_i - w_j - k_j\, r$$

In figure 11.3 expressions (13) are shown as straight lines with slopes $k_j\, b_j$ equal to transportation costs of these outputs from one unit of labor on one unit of land and intercepts $p_j b_j$ the product prices received in the market for that output.

Observe that one product, flax, is dominated by the others and therefore not grown anywhere. Notice that once more products are grown in the sequence of decreasing weights (transportation costs) $k_j\, b_j$ as one moves away from the center.

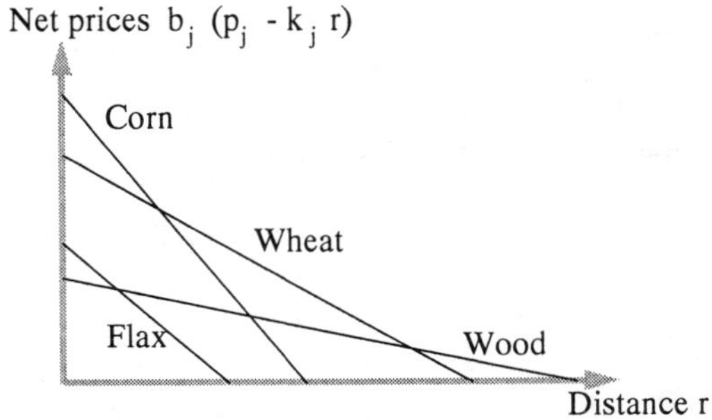


**Fig. 11.3.** How received prices decline with distance for outputs from unit inputs

## Mathematical Note 11.4
## The Closed von Thünen Economy

We consider a closed economy (the isolated state) with three sectors: agriculture, manufacturing and transportation. Their technologies are described by Cobb Douglas functions as follows:

Agriculture output q per land area results from labor input x per land area as

$$q = x^{\alpha} \tag{1}$$

Let transportation cost k for the agricultural product be standardized at k = 1. Prices received by farmers located at distance r from the center are then

$$p - r.$$

Wage rates are standardized at w = 1. Profit maximization by farmers results in decisions

$$\underset{x}{\text{Max}}\,(p-r)x^{\alpha} - x$$

$$0 = \alpha(p-r)x^{\alpha-1} - 1$$

$$x = \left[\alpha(p-r)\right]^{\frac{1}{1-\alpha}} = c_0(p-r)^{\frac{1}{\beta}} \qquad \text{(say)} \tag{2}$$

where

$$c_0 = \alpha^{\frac{1}{1-\alpha}} \qquad \beta = 1 - \alpha$$

Output per land at distance r is

$$q = x^{\alpha}$$

and aggregate agricultural output is

$$Q_A = \int_0^R q 2\pi r dr$$

Aggregate labor input in agriculture is

$$X_A = \int_0^R x2\pi r dr$$

Industrial production located in the city using aggregate labor $X_I$ produces aggregate output

$$Q_I = \beta x_I^\gamma \qquad 0 < \gamma \leq 1 \tag{3}$$

at constant or diminishing returns (to labor). When capital inputs – not considered here – are also introduced the combined inputs may have increasing returns to scale.

Transportation is assumed to require only labor in fixed proportion to transport output, standardized at unity

$$X_T = T = \int_0^R qr \cdot 2\pi r dr \tag{4}$$

The industrial product is shipped free as back freight from the production site in the metropolis to consumers in the country and is thus not included.

The available labor force N (proportional to total population) must then be allocated subject to a labor availability constraint

$$X_A + X_I + X_T \leq N \tag{5}$$

Under perfect competition this allocation results in a welfare maximum. We assume that all consumers have identical utility functions in terms of the two commodities: agricultural and industrial product

$$u = \lambda \log q_A + \mu \log q_I$$

where $q_A$ and $q_I$ denote individual consumption. Welfare in terms of aggregate consumption $Q_A$, $Q_I$ is then

$$U = \lambda \log Q_A + \mu \log Q_I \tag{6}$$

We now calculate the quantities produced and consumed in equilibrium and the amounts of labor used. It turns out that the equilibrating variable is the price p of the agricultural product in the metropolis which – because of our standardization k = 1 – is also the radius R of the land area used for agriculture

$$p = R \tag{7}$$

Aggregating agricultural production

$$Q_A = \int_0^R c_0^\alpha x^\alpha 2\pi r dr$$

$$= 2\pi c_0^\alpha \int_0^p (p - r)^{\frac{\alpha}{\beta}} r dr$$

$$= \frac{2\pi c_0^{\alpha}}{\left(1+\frac{\alpha}{\beta}\right)\left(2+\frac{\alpha}{\beta}\right)} p^{\frac{\alpha}{\beta}+2}$$

$$= c_1 p^{\frac{\alpha}{\beta}+2} \qquad \text{(say)} \tag{8}$$

Aggregating agricultural labor

$$X_A = \int_0^R x 2\pi r dr$$

$$= 2\pi c_0 \int_0^p (p-r)^{\frac{1}{\beta}} r dr$$

$$= \frac{2\pi c_0}{(1+\beta)(2+\beta)} p^{\frac{1}{\beta}+2} = c_2 p^{\frac{1}{\beta}+2} \tag{9}$$

Calculating transportation requirements

$$T = \int_0^R x^{\alpha} 2\pi r^2 dr$$

$$= 2\pi c_0^{\alpha} \int_0^p (p-r)^{\frac{\alpha}{\beta}} r^2 dr$$

$$= \frac{2\pi c_0^{\alpha}}{(1+\alpha/\beta)(2+\alpha/\beta)(3+\alpha/\beta)} p^{\frac{\alpha}{\beta}+3} = c_3 p^{\frac{\alpha}{\beta}+3} \tag{10}$$

Observe now that

$$\frac{\alpha}{\beta}+3 = \frac{\alpha}{1-\alpha}+3 = \frac{1}{1-\alpha}+2 = \frac{1}{\beta}+2$$

so that labor input into transportation turns out to be proportional to agricultural labor

$$X_T = T = c_3 p^{\frac{1}{\beta}+2} = \frac{c_3}{c_2} \cdot X_A \tag{11}$$

Using the budget constraint, industrial labor is thus

$$X_I = N - \left((c_2 + c_3)p^{\frac{1}{\beta}+2}\right)$$

Welfare is now

$$W = \lambda \log c_1 p^{\frac{1}{\beta}+1} + \mu\gamma \log\left[N - (c_2 + c_3) p^{\frac{1}{\beta}+2}\right]$$

Maximization with respect to p, which governs labor allocation, yields

$$0 = \frac{\lambda\left(1+\frac{1}{\beta}\right)}{p} - \frac{\mu\gamma \cdot (c_2 + c_3)\left(2+\frac{1}{\beta}\right)p}{N - (c_2 + c_3)p^{2+\beta}}$$

$$\mu\gamma(c_2 + c_3)(2+\beta)p^{2+\frac{1}{\beta}} = \lambda(1+\frac{1}{\beta})\left[N - (c_2 + c_3)p^{2+\frac{1}{\beta}}\right]$$

$$p^{2+\frac{1}{\beta}} = \frac{\lambda\left(1+\frac{1}{\beta}\right)N}{\mu\gamma(c_2 + c_3)\left(2+\frac{1}{\beta}\right) + \lambda\left(1+\frac{1}{\beta}\right)(c_2 + c_3)}$$

so that

$$R = p = c_4 N^{\frac{1}{2+\frac{1}{\beta}}} \tag{12}$$

This shows how the radius of agricultural land expands with increasing population.

Agricultural labor

$$X_A = c_2 p^{2+\frac{1}{\beta}} = c_2 c_4 N^{2+\frac{1}{\beta}} \tag{13}$$

takes up a fixed proportion of the total labor force and so do transportation and industry. (This is a result of the log linear specifications of production and utility functions). Shifts of labor proportions into industry can result here only through technical progress.

## Mathematical Note 11.5
## The Monocentric City

Consider a utility function u = u (s, c)

s space
c general consumption
y income
p housing rent per square meter
k communing cost per kilometer
r distance from the city center

The budget equation states

$$c = y - ps - kr \tag{1}$$

Specify $u = a \ln s + \ln c$

Utility maximization

$$\underset{r,s}{\text{Max}}\ a \ln s + \ln\left(y - ps - kr\right) \tag{2}$$

with respect to space

$$0 = \frac{\partial}{\partial s} = \frac{a}{s} - \frac{p}{y - ps - kr}$$

$$ps = \frac{a}{1+a}\left(y - kr\right) \tag{3}$$

Housing expenditure is proportional to income net of commuting cost ≈ proportional to income

The achieved utility is

$$u = a \ln \frac{a}{1+a}\frac{y-kr}{p} + \ln \frac{1}{1+a}\left(y - kr\right)$$

$$= a \ln a - \left(1+a\right)\ln\left(1+a\right)$$

$$+\left(1+a\right)\ln\left(y-kr\right) - a \ln p$$

In all locations where households choose to live this achieved utility must be the same, and the same as at distance $r = 0$ adjacent to the central business district. This determines $p = p(r)$

$$a \ln a - \left(1+a\right)\ln\left(1+a\right) + \left(1+a\right)\ln\left(y-kr\right) - a \ln p\left(r\right)$$

$$= a \ln a - \left(1+a\right)\ln\left(1+a\right) + \left(1+a\right)\ln y - a \ln p(0)$$

$$\ln p\left(r\right) = -\frac{1+a}{a}\ln y + \frac{1+a}{a}\ln\left(y-kr\right) + \ln p\left(0\right)$$

$$\ln p\left(r\right) + \ln p(0)\frac{1+a}{a}\lambda v\left(1 - \frac{k}{y}r\right) + \ln p(0)$$

$$p(r) = p\left(0\right)\left[1 - \frac{k}{y}r\right]^{\frac{1+a}{a}} \tag{4}$$

showing rent to be a convex function of distance r decreasing to zero at a critical distance $R = \frac{y}{k}$ when all income would be spent on commuting.

With the approximation

$$e^{-x} \doteq 1 - x \tag{5}$$

this becomes

$$p(r) \doteq p(0)e^{-\frac{1+a}{a}\frac{k}{y}r} \tag{6}$$

Housing space demanded per household is then, using (3) in (4)

$$s(r) = \frac{a}{1+a} \cdot \frac{y-kr}{p} = \frac{a\,y^{\frac{1+a}{a}}}{(1+a)p(0)}(y-kr)^{-\frac{1}{a}}$$

Residential density, the inverse of space per household, is then

$$\frac{1}{s} = \frac{(1+a)p(0)}{a} \cdot \left(1-\frac{kr}{y}\right)^{\frac{1}{a}} \tag{7}$$

or with the approximation (5)

$$d(r) \doteq d(0)e^{-\frac{k}{y}r} \tag{8}$$

an exponential decline.

### Mathematical Note 11.6
### Land Use under Spatially Dispersed Demand

In the von Thünen model all land use is oriented towards a given market in the metropolis, a predetermined fixed point in space. What happens when demand is spatially dispersed?

For simplicity we assume an elementary case: products i = 1, ... m for which the density of demand $q_i(x)$ (demand per unit area) is given at all locations $x = (x_1, x_2)$.

We use the "efficient market hypothesis" that under perfect competition, total cost of production and transportation incurred to meet this demand is minimized. We also assume constant returns to scale in the independent production functions $f_i$ and land and labor as the only factors of production. Let labor input per unit of land used – the labor intensity of land use – be $l_i$ and let $m_i$ be the proportion of an area unit allocated to the production of i. Output of i per area – its density – is then

$$m_i f_i(l_i).$$

The area use is constrained by

$$\sum_i m_i \le 1 \tag{1}$$

In addition there is a commodity balance equation

$$\text{local output} = \text{local consumption} + \text{net exports}$$
$$m_i f_i = q_i + \operatorname{div}\phi_i \tag{2}$$

where $\phi_i$ denotes the vector of commodity flow. Incorporating these constraints in the efficient market hypothesis leads to

$$\max_{l_i m_i \phi_i} \iint \left( -wi\sum_i l_i m_i - \sum_i k_i |\phi_i| + \rho \left[ 1 - \sum m_i \right] \right.$$

$$\left. + \sum_i \lambda_i \left[ m_i f_i(l_i) - q_i - \operatorname{div}\phi_i \right] dx \ dx \right. \tag{3}$$

The Euler-Lagrange conditions for a maximum are then

$$m_i \begin{Bmatrix} = \\ \geq \end{Bmatrix} 0 \Leftrightarrow \lambda_i f_i(l_i) - wl_i \begin{Bmatrix} < \\ = \end{Bmatrix} \rho \tag{4}$$

$$f_i' = w \tag{5}$$

$$\phi_i = 0 \quad \text{if} \quad |\operatorname{grad}\lambda_i| < k_i$$

$$k_i \frac{\phi_i}{|\phi_i|} = \operatorname{grad}\lambda_i \quad \text{if} \quad \phi_i \neq 0. \tag{6}$$

Condition (5) is the well-known principle of marginal productivity applied to labor. Condition (6) states the – less well-known – equation for the efficient flow direction. Equation and inequality (4) states that under constant returns to scale profits (per unit of land) are absorbed as land rent $\rho$. More importantly (4) implies

$$\rho = \max_i \left[ \lambda_i f_i - wl_i \right] \tag{7}$$

Land is allocated exclusively to the most profitable land uses.

The generalization of von Thünen rings are now areas of specialized land use whose pattern will depend on both supply – the productivity of land for particular products – and on demand: the spatial distribution of consumption $q_i(x)$.

For a centrally oriented flow field the conditions (4), (5), (6) reduce to the equations of the von Thünen model.

## References

Alonso W (1964) Location and Land Use. Harvard University Press, Cambridge

Beckmann JJ (1968) Location Theory. Random House, New York

Beckmann MJ (1969) On the Distribution of Urban Rent and Residential Density. Journal of Economic Theory, Vol 1, No 1, June, pp 60–67

Beckmann MJ (1972) Von Thünen Revisited: A Neoclassical Land Use Model. Swedish Journal of Economics 74:1–7

Beckmann MJ (1976) Spatial Equilibrium in the Dispersed City. In: Papageorgiou YY (ed) Mathematical Land Use Theory. Lexington Books, Lexington, Ma., p 117–125

Beckmann MJ, Puu T (1985) Spatial Economics: Flow, Density and Potential, North-Holland, Amsterdam

Clark C (1945) The economic function of a city in relation to its sice. Econometrics 13:97–113

Dunn ES (1954) The Location of Agricultural Production. Gainesville

Muth RF (1969) Cities and Housing. University of Chicago Press, Chicago

Solow R, Vickrey WS (1971) Land Use in a Long Narrow City. Journal of Economc Theory 3:430–337

Steuart J (1966) An Inquiry into the Principles of Political Economy. Reprint. University of Chicago Press, Chicago

von Thünen JH (1826) Der isolirte Staat in Beziehung auf Landwirtschaft und Nationalökonomie. Stuttgart. English translation: The Isolated State (1966). Pergamon Press, Oxford

# CHAPTER 12

# Urban Systems

Human settlements have been founded for various reasons: as seats of government and the administration of the national territory and its subdivisions, for worship, for defense, for medical cures and for recreation and sports. The economic role of settlements has been that of land cultivation, of resource extraction, manufacturing, and in the case of settlements qualifying as cities, of markets.

At early stages of economic development when population densities and incomes are low, markets are held only periodically. Later cities became permanent market sites.

The urban system of a nation is not always the result of purely economic forces but may be anchored to a set of administrative centers, located by government decisions. There will be a national capital city (which need not be the economic capital, but in older nations it often is). Next there are regional centers, provincial or state capitals, district centers, county seats or similarly designated locations of a hierarchically organized administration, where ranks are assigned to centers in accordance with the rank system of the administrative organization.

Central places are defined as centers having market functions as well as offering administrative services. Historically, either function may be located first and then attract the other. Government administration is hierarchically organized and the administrative centers are ranked from the national capital through state, provincial and regional centers down to the county level or equivalent. Market places of various size can exist well below the county level. The question is, whether economic services, i.e. market functions, will also fall into a hierarchical pattern?

If so, one speaks of a *central place system.*

A strict ranking of central places can be recognized when high ranking central places offer all services of lower ranking ones plus some more. This is the outcome of economies of scale and a minimum market size needed for reaping such economies. They will differ for different products and services. Products and services become available in centers that afford this market size locally and in the region served by the center.

In addition to trading, central places will also attract market oriented industries. The size of the market required is then shown by the profit function (3.3.19) to be increasing with fixed cost. For survival the spread between demand intercept a and unit production cost c adjusted for transportation cost

$$\frac{a-c}{k}$$

must then also be large enough to offset the increasing average distance over which the good must be transported (mathematical note 12.1). The locating together of plants from different industries in centers that offer the required minimum market size is an example of agglomeration not resulting from interaction. A ranking of products then emerges from the ranking of places thus translates, defined by the lowest place rank at which this good is offered.

Since all cities of the same rank offer the same range of goods and services there is no need for trade among them. All trade is in a vertical direction between higher and lower ranking places: those of higher rank selling higher ranking goods (services) to lower ranking places and receiving agricultural goods in exchange.

When resource oriented industries are included, particularly those processing agricultural products, the central place system is no longer strictly hierarchical. For processing of such agricultural materials as sugar cane or sugar beets, fruits, vegetables and milk, occur near the agricultural basis in low ranking centers, that otherwise only supply basic services.

New industries are best launched from the metropolis, the center of highest rank, since it offers access to the largest market, both locally and regionally. As long as $\frac{a-c}{k} < \bar{r}_N$ where $\bar{r}_N$ is average distance to the center in the entire region, the product market is essentially local, i.e. the metropolis itself. As demand develops, the rising $\frac{a-c}{k}$ allows serving more distant points and permits the entry of more firms in the metropolis as well as decentralization to cities of the next lower ranks.

Centers of the lowest order may be not more than a crossing point of highways offering a gas station, convenience store and motel. At the next rank may be added: school, church, food store, drug store and a repair shop for automobiles and/or agricultural machinery. In North America, the smallest distribution centers have all declined in the range of goods and services offered and in population as well, due to the greater mobility of rural customers. They now prefer the wider choice and better prices at higher order central places, which they can reach conveniently by automobile (Mc Pherson 1973).

Examples of central places have been discovered in many regions of the world: South Germany, the original example of Walter Christaller (1933) the Mississippi valley of North America and the Punjab of India, to name but a few (Berry and Pred 1961). Most urban systems appear to be a mixture of central places and industrial centers having resource oriented industries. This would create some regional specialization and thus trade among cities of the same rank (assuming rank can still be assigned in a meaningful way).

The set of industries that locates together in one center need not have the same composition, as demonstrated in figure 9.1 of chapter 9 for a linearly extended economy. However, the following principle must hold. Since a settlement of rank n also serves as market center for one of rank n-1, those activities that are common to all centers of rank n-1 also occur in those of rank n.

In this approach to urban systems we have, following Lösch (1940), built up a system of market centers of increasing size and complexity from the bottom up, starting with the activities that require the smallest market areas and were the subject of chapter 7 on the location of a dispersed industry.

Starting from the top down one can derive a hierarchical system of urban places as seats of the executive (administrative) and service activities of government as was done by Christaller (1934). In either case one obtains a "central place system" as the set of locations for services and market oriented industries.

The size of a central place is determined by the set of economic activities it supports. This apparently quite complicated relationship can be approximated by a simple principle of proportionality. Suppose that the population of a city serving its market area is proportional to that of the market area population served. Assume further that the number m of centers of the next lower rank from which one center of rank n was selected is the same at all levels. Then both city size and population of market areas increase exponentially at the same rate with rank. If we take the city of medium size in a rank class and count its place in descending order among all cities arranged by size, then the famous "rank-size rule" (Zipf 1941) results (Beckmann 1958). The two numbers are in a decreasing log linear relationship to each other (figure 12.3). This is shown in mathematical note 12.3.

An alternative approach to size in urban systems formulated in term of income and expenditure is that of Tinbergen and Bos (1959), (Paelinck 1966).

**Fig. 12.3.** Rank-size rule

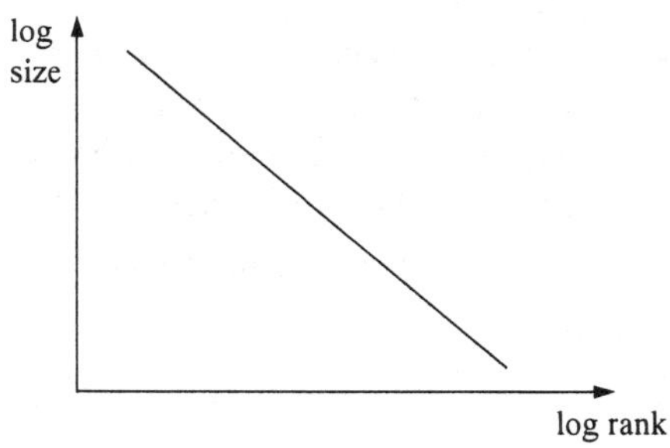


### Mathematical Note 12.1
### Market Size for Central Place Goods

To show the required market size $M = M(R)$ write the profit function for (either mill or) uniform pricing, (recalling 7.1.1) in the form

$$\left[a - c - k\bar{r}(R)\right]^2 M(R) \geq 4F \qquad (1)$$

to see that

- higher ranking commodities have larger fixed cost F
- therefore the population served $M(R)$ must be larger
- to offset the larger average distance $\bar{r}(R)$ in the regional market, the standard coefficient $\frac{a-c}{k}$ must also be sufficiently large.

For every market oriented good, a critical R can be calculated such that M(R) and $\bar{r}(R)$ satisfy condition 1. The good is then offered in all central places whose market size equals or exceeds M(R).

Observe that population density is not constant in the market area of a central place, for this market area contains centers of all lower ranks. The simple profit formulae based on constant population density therefore do not apply here.

**Mathematical Note 12.2**
**Basic Grid Length in Hexagonal Tiling**

The basic grid of points for human settlements is supplied by the locations of that industry (or economic activity) that (in equilibrium) needs the smallest market areas. If these market areas are hexagonal then the next larger hexagons whose centers can fall onto those of the basic grid must have a market radius (of the inscribed circle) of two or $\frac{3 \cdot 2}{\sqrt{3}} = 2\sqrt{3}$ times the radius in the basic grid (figure 12.1).

For a Manhattan metric the market radii are n $R_0$ with n = 2, 3, 4, 5... . Anyway, as the market radius goes up by a factor $\lambda$, the market area increases by $\lambda^2$ and the number of firms or settlements goes down by the factor $\lambda^2$.

No matter what the geometry of the basic set of the smallest settlements is – even a random grid – the larger market areas cannot increase continuously but only in discrete steps if the larger centers are restricted to points of the basic grid. The same applies at every stage. If one out of m grid points at level n becomes a center at level n + 1, then the market area increase by m and the distances by $\sqrt{m}$.

Fig. 12.1

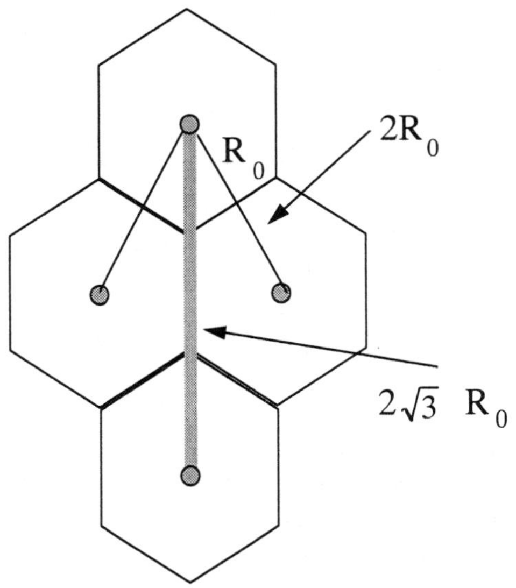

### Mathematical note 12.3
### Derivation of the Rank-size Rule

Let

$c_n$ average size of centers of rank n

$M_n$ population of market area of a center of rank n

We postulate

$$c_n = bM_n \tag{2}$$

A cluster of rank n serves m − 1 centers of rank n − 1 and m market areas of rank n − 1.

The population of $M_n$ is composed of m populations $M_{n-1}$ plus that of the center $c_n$ which will be larger by $c_n - c_{n-1}$. Therefore

$$M_n = mM_{n-1} + c_n - c_{n-1} \tag{3}$$

Using (2) to eliminate $M_n$ from (3)

$$\frac{c_n}{b} = \frac{m}{b}c_{n-1} + c_n - c_{n-1}$$

$$c_n = \frac{\left(\frac{m}{b} - 1\right)}{\frac{1}{b} - 1} c_{n-1}$$

$$c_n = B \cdot c_{n-1} \qquad \text{(say)}$$

This difference equation has the solution

$$c_n = c_0 B^n \tag{4}$$

where $c_0$ is the size of the smallest settlements. When ordered in decreasing size the median city in rank class n is in place

$$\nu_n = 1 + m + \ldots + m^{N-n-1} + \frac{m^{N-n}}{2}$$

$$= \frac{m^{N-n} - 1}{m - 1} + \frac{m^{N-n}}{2}$$

$$= A \cdot m^{N-m} - \frac{1}{m-1} \qquad \text{(say)}$$

Neglect $\frac{1}{m-1}$ which is small compared to the first term

$$\nu_n \doteq A \cdot m^{N-m} \tag{5}$$

Taking logarithms in (4) and (5) and writing $\beta = \log B \quad \alpha = \log m$

$$\log c_n = \log c_0 + \beta n$$

$$\log \nu_n = \log A + (N - n)\log m$$

Eliminating n

$$\log \nu_n = \text{cons tan t} - \frac{\alpha}{\beta}\log c_n \tag{6}$$

In the rank size rule the slope $\frac{\alpha}{\beta}$ is found empirically to be close to unity

or

$$\log m = \alpha \doteq \beta = \log \frac{\frac{m}{b} - 1}{\frac{1}{b} - 1} = \log \frac{m - b}{1 - b} \tag{7}$$

This is true when the proportionality factor $b < 1$ is small compared to m.

Thus for $m = 4$, $b = \frac{1}{10}$

$$\beta = \ln \frac{3{,}9}{0{,}9} = \ln 4{,}33 = 1.46$$

$$\alpha = \ln 4 = 1.386$$

$$\frac{\alpha}{\beta} = 0.945$$

## References


Beckmann MJ (1958) City Hierarchies and the Distribution of City Sizes. Economic Development and Cultural Change VII, No 3:243–248

Beckmann MJ, McPherson J (1970) City Size Distribution in a Central Place Hierarchy: An Alternative Approach. Journal of Regional Science 10, No. 1:25–33

Berry B, Pred A (1961) Central Place Studies: A Bibliography of Theory and Applications. Regional Science Research Institute, Philadelphia

Bos HC (1965) Spatial Dispersion of Economic Activity. Rotterdam University Press, Rotterdam

Christaller W (1933) Die zentralen Orte in Süddeutschland. Jena

Eaton BC, Lipsey RG (1982) An Economic Theory of Central Places. Economic Journal 92:56–72

Henderson JV (1972) Hierarchy Models of City Size: An Economic Evaluation. Journal of Regional Science 12:435–441

Hendersson JV (1974) The Sizes and Types of Cities. American Economic Review 64:640–656

Mc Pherson J (1973) Economics of Central Place Hierarchies. Application to the Great Plains Experience. PhD Thesis, Brown University

Paelinck JHP (1966) Empirical Evidence on Tinbergen-Bos Systems. Center of Economic Studies, University of Munich, Germany, January–May

Stewart JQ (1947) Empirical Mathematical Rules Governing the Distribution and Equilibrium of Population. Geographical Review 48:222–245

Tinbergen J (1959) Selected Papers. North-Holland, Amsterdam

Zipf GK (1941) National Unity and Disunity. Principia Press, Bloomington, IN

CHAPTER 13

# Transportation Systems

In the short run a transportation network is in place and the question is: how its capacity should be utilized. In the long run we are interested in the structure and location of the transportation networks.

A distinction must be made between the operation of public transportation systems and the traffic flows that are the results of individual decisions by operators of private vehicles. In our society the provision of transport facilities other than vehicles, that is the transportation network, is a public service of government.

As a start consider the transportation network with the simplest structure: air transport. It is defined by points only: origins, destinations and nodes. At its most elementary an air transport network consists of one central node: the hub and the remaining origins and destinations at the end of spokes issuing from that hub. Originations and terminations of passengers (demand accommodated) does then uniquely determine all passenger flows and these define the required flows of plane capacities. The only allocation decision is then, what demand should be accommodated. When all routes are profitable and plane capacity is sufficient, this means: all demand. Otherwise the demand with the lowest profit margins will be left unserved. Since traversing the hub usually increases distance above a direct connection, direct flights between origin destination pairs will be more profitable when demand suffices for an adequate load factor. Another decision problem concerns the best use of planes of different capacities and operating costs. Economies of scale exist such that transportation cost per passenger decreases with the size of the plane, assuming constant load factors.

More interesting for economic analysis is the allocation of passenger traffic among hubs in a multi-hub system. In the absence of capacity restrictions for hubs, the one offering the shortest flight paths would always be chosen for any given origin destination pair. When capacities are limited, the choice can be guided by efficiency prices to be imposed on the use of a hub with excess demand. These efficiency prices should become market prices discovered through the auctioning off of slots at hubs to competing carriers. As it is, hubs are often the property of certain airlines who can ward off unwanted competition by reserving slots for their own use exclusively.

So far demand has been assumed given. We can still use linear programming models for finding optimal allocations when quantity demanded is considered a step function of fares, and fares are calculated at marginal cost. This ignores all fixed cost. A welfare optimizing way to recover fixed cost would be to charge a club membership fee to all potential users covering these fixed costs. Otherwise a certain amount or percentage would have to be added to marginal cost, an inferior but common method for financing fixed costs.

The next level of complication in the utilization of transport networks arises when capacity limits exist also on the links between points (nodes) as well as in the nodes themselves. This is characteristic of railroad transportation. Railroad hubs have existed for a long time. In addition to efficiency prices or rents for the use of congested intersections or hubs one now needs prices for link capacities as well.

In highway transportation and road transport within urban areas the relationship between traffic load and capacity is a more complicated than a fixed limit on the flow that can be passed through. In fact as traffic increases, passage time increases going to extreme levels as the capacity limit is reached. The exact details are of a technological nature examined in "Traffic science" (Herman 1961). As an approximation, travel time is often considered a quartic function of the difference between capacity and actual flow.

$$t = a \cdot (c - x)^4$$

In choosing among alternative routes individual traffic attempts to avoid the worst bottlenecks, but does not do so to a sufficient extent. For private cost – the travel time experienced individually – is short of social cost – which includes the marginal delays imposed on others. To economists the obvious remedy would once more be pricing, in this case charging the difference between social and private costs as a user fee. But so far road pricing – for the purpose of controlling congestion rather than financing special toll roads – has proved politically unpopular, except in that modern state of enlightened despotism: Singapore.

The study of traffic flow on road networks involves thus two distinct questions: prediction and optimization. Prediction of traffic resulting from individual road user decisions unguided by road pricing, requires the analysis of an equilibrium resulting from the interaction of these choices with the available (non-linear) capacities (math.note 13.2).

Optimization requires the determination of efficiency prices, or of the desired traffic flows directly while leaving open the question of how to bring about these flows, when some road users would prefer other routes as better for themselves.

The location of transportation networks raises interesting problems of various degrees of difficulty. Consider the simplest case: linking three given points. A shortest network results as the solution of a Weber problem with the expected traffic loads as the appropriate weights pulling the intersection of the three link roads into a cost minimizing position. When road construction costs per mile and traffic flows are equal on all links of the network, the location of a cost minimizing road network reduces to a "Steiner problem", a mathematical classic of known complexity. The problem of locating a hub is a straight forward generalization of this, except that hubs must be located in or near a major origin/destination point. As demands vary one should anticipate discontinuous adjustments of that hub location. The locations of secondary hubs are even less stable and also more difficult to determine.

The spatial structure of a transportation system that links a given set of cities which generate a known demand for inter urban transportation is a complex problem of integer programming, to whose solution location theory cannot contribute much. A few general principles can be stated.

- *Hierarchy.* Economics of scale mandate that long distance transportation should be bundled to flows in a network of super highways from which access to particular destinations is gained through a system of regional and finally local highways.
- *Capacity.* The supply of capacity should respond to the demand for traffic. There is thus a case for multi lane and limited access facilities issuing from major traffic generating centers.
- *Centrality.* The optimal number of spokes in a traffic hub should be an increasing function of the total traffic volume at a given hub.

## Mathematical Note 13.1
## Traffic Flow in Networks

A transportation network consists of links ij joining nodes i, j in which traffic is generated or terminated or passing through. The network is considered to be a connected graph (figure 13.1).

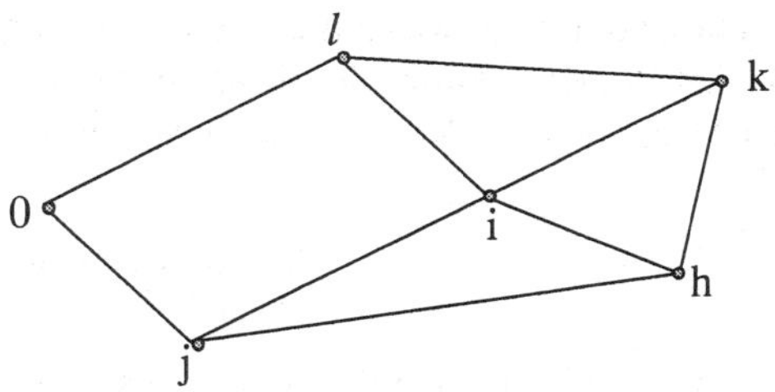


**Fig. 13.1.** A simple transportation network

To begin with, let all traffic be directed to a single destination i = 0. At all other locations $i \neq 0$, let $q_i$ denote traffic originating. Traffic flows $x_{ij}$ must then obey a conservation law (first formulated by Kirchhoff for electric currents).

$$\sum_j x_{ij} - x_{ji} = q_i \tag{1}$$

where the summation extends only over pairs ij representing links in the network. Equation (1) applies also to destination i = 0 if we define

$$q_0 = -\sum_{i \neq 0} q_i \tag{2}$$

so that, in fact, all originations add up to zero.

Presumably travelers who make up this traffic intend to reach their destination by a shortest path. When link lengths $r_{ij}$, which eventually determine path lengths to destination 0, are in units of time, the traveler seeks a path of least time. Both geographical distance and time may be considered specifications of cost. Thus the objective of travelers' choices may be defined as least cost, although for convenience we shall term it a shortest path.

Finding shortest paths, if not by visual inspection of a map, can be formalized as a linear program (below) or more intuitively as a dynamic program.

A dynamic program for this purpose is generated as follows. At destination 0 the path length is zero

$$D(0) = 0 \tag{3}$$

Let $D_n(j)$ denote path lengths from j by shortest routes involving no more than n links. When more links are needed, define

$$D_n(j) = \infty$$

or

$$D_n(j) = A \qquad A >> 1 \tag{4}$$

where A is a large number, exceeding e.g. the sum of all links lengths $r_{ij}$

$$A > \sum_{ij \in N} r_{ij} \tag{5}$$

Now let $D_n(0) = 0$ for all n

Let i be a point from which it is possible to reach destination 0 by a path of at most n links. Let $D_{n-1}$ (j) be known for those neighbor points j of i that are at most n-1 links removed from point 0. A shortest path with at most n links from i is then determined as

$$D_n(i) = \min_j \left[ r_{ij} + D_{n-1}(j) \right] \tag{6}$$

As n is increased it relaxes the restrictions on intermediate points of a path so that $D_n(i)$ is a non-increasing function of n. Going to the limit in (6) the unrestricted shortest path lengths D(i), D(j) are generated as solutions of the Bellman equation (6) satisfying

$$D(i) = \min_j \left[ r_{ij} + D(j) \right] \tag{7}$$

$$D(0) = 0.$$

The shortest paths is then the sequence of intermediate points j that solve (7).

The same shortest path now emerge from a linear program in terms of the actual flows $x_{ij}$ on link ij. The object is to minimize total transportation costs

$$\min_{x_{ij} \geq 0} \sum_{ij} r_{ij} x_{ij} \tag{8}$$

subject to

$$\sum_j \left( x_{ij} - x_{ji} \right) = q_i \tag{1}$$

yielding the Koopmans efficiency conditions

$$x_{ij}\begin{Bmatrix}=\\ \geq\end{Bmatrix}0 \Leftrightarrow \lambda_i - \lambda_j\begin{Bmatrix}<\\ =\end{Bmatrix}r_{ij} \tag{9}$$

Here the efficient prices or Lagrange multipliers $\lambda_i$ represent distance to destination

$$\lambda_i = D_i$$

provided we set

$$\lambda_0 = 0 \tag{3}$$

as we may since the $\lambda_i$ are determined only up to an additive constant. For (9) implies

$$\lambda_i = \min_j\left[r_{ij} + \lambda_j\right] \tag{6}$$

The efficiency condition (9), (3) together with the flow balance equation (1), (2) jointly determine the traffic $x_{ij}$ on the links ij.

When several destinations k are considered (in fact all nodes of the network are potential destinations) it is merely necessary to index flows by their destination $x_{ij}^k$ (say)

Travel demand is in terms of both origins i and destination k, $q_i^k$ and condition (1) is modified

$$q_k^k = \sum x_{ij}^k - x_{ji}^k \tag{1a}$$

The distance functions now carry the destinations k

$$\lambda_i^k = \min_j\left(r_{ij} + \lambda_j^k\right) \tag{6}$$

and total flow $x_{ij}$ on link ij

$$x_{ij} = \sum_k x_{ij}^k \tag{10}$$

is the sum of the individual flows.

The next step is to introduce capacity constraints on links $x_{ij}$, or transit points j

$$x_{ij} \leq c_{ij} \tag{11}$$

$$\sum_j\left(x_{ij} + x_{ji} \leq c_i\right)$$

These constraints will generate efficiency prices $\mu_{ij}$, $\gamma_i$ that augment the costs $r_{ij}$ to travelers on link ij so that

$$r_{ij} + \mu_{ij} + \gamma_i + \gamma_j \tag{12}$$

becomes the effective cost of using link ij, thereby inducing detours.

These charges $\mu_{ij}$, $\gamma_i$ will vanish when capacities are under utilized

$$\mu_{ij}\begin{Bmatrix}=\\ \geq\end{Bmatrix}0 \Leftrightarrow x_{ij}\begin{Bmatrix}<\\ =\end{Bmatrix}c_{ij} \tag{13}$$

$$\gamma_i\begin{Bmatrix}=\\ >\end{Bmatrix}0 \Leftrightarrow \sum_j\left(x_{ij}+x_{ji}\right)\begin{Bmatrix}<\\ =\end{Bmatrix}c_i \tag{14}$$

Capacity limits on links and nodes also raise problems of feasibility: not all demand can be accommodated when capacities are too low. For a single destination 0 it can be shown that (Ford-Fulkerson 1962) max flow = min cut where cut is the sum of capacities of all links intersected by a closed curve around the destination.

**Mathematical Note 13.2**
**Traffic Equilibrium**

In a more sophisticated approach to private vehicle traffic in networks, the capacity of the roads (links) is replaced by functions that show a relationship between travel time $t_{ij}$ and flow volume $x_{ij}$ on that link

$$t_{ij} = t_{ij}\left(x_{ij}\right) \tag{15}$$

The optimal utilization of a road network to accommodate given demands $q_i$ for trips to a single destination 0 is then found by solving a convex nonlinear program

$$\underset{x_{ij}\geq 0}{\text{Min}} \sum_{ij} t_{ij}\left(x_{ij}\right)x_{ij} \tag{16}$$

$$\sum_j x_{ij} - x_{ji} = q_i \tag{1}$$

whose Kuhn-Tucker conditions are

$$x_{ij}\begin{Bmatrix}=\\ \geq\end{Bmatrix}0 \Leftrightarrow \lambda_i - \lambda_j\begin{Bmatrix}<\\ =\end{Bmatrix}t_{ij} + x_{ij}t'\left(x_{ij}\right) \tag{17}$$

Here the travel cost (or transportation time) $t_{ij}$ experienced by the driver must be augmented by a user charge

$$x_{ij}t'\left(x_{ij}\right)$$

that represents the marginal delays inflicted by the vehicle on all the traffic on this road.

To model the actual traffic equilibrium without this user charge one considers

$$\underset{x_{ij}\geq 0}{\text{Min}} \sum_{ij} \int_0^{x_{ij}} t_{ij}(x)dx \tag{18}$$

subject to

$$\sum_j x_{ij} - x_{ji} = q_i \qquad (1)$$

which yields

$$x_{ij} \begin{Bmatrix} = \\ \geq \end{Bmatrix} 0 \Leftrightarrow \lambda_i - \lambda_j \begin{Bmatrix} = \\ < \end{Bmatrix} t_{ij} \qquad (19)$$

as in (13.1.9) before. User charges depending on traffic volume are popular only among economic theorists. It seems that things have to get worse before they can get better. Notice that expression (18) has no economic meaning indicating that traffic equilibrium without user charges is a non-optimum.

## Mathematical Note 13.3 Allocation in Public Transportation

Once more let $q_i^k$ demand for trips from i to k

and $q_k^k = -\sum_{i \neq k} q_i^k$

for trips terminating in k.

More concretely write

$x_{ij}^k$ travelers to k on link ij
$v_{ij}$ vehicle flow on link ij going towards j
a vehicle capacity
$k_{ij}$ transportation cost of one vehicle on link ij

We assume steady state flows, e.g. at peak hours or peak days, and ignore scheduling times.

The piecing together of complete vehicle routes from link trips $v_{ij}$ and their scheduling – apart from frequencies – will not be considered here. Any delays or inconvenience to travelers from switching to different vehicles along their routes are considered small enough to be ignored.

At first consider the object of public transportation to be the minimizing of total transportation costs incurred by the public authority

$$\min_{v_{ij}, x_{ij}^k \geq 0} \sum_{ij} k_{ij} v_{ij} \qquad (1)$$

subject to the requirement of meeting demand

$$a v_{ij} \geq \sum_k x_{ij}^k \qquad (2)$$

$$\sum_j \left( x_{ij}^k - x_{ji}^k \right) = q_i^k \qquad (3)$$

a linear program in flow variables $v_{ij}$, $x_{ij}^k$.

Consider the Lagrangean

$$L = -\sum_{ij} k_{ij} v_{ij} + \sum_{ij} \mu_{ij} \left( a v_{ij} - \sum_k x_{ij}^k \right)$$

$$+ \sum_{i,k} \lambda_i^k \cdot \left[ \sum_j \left( x_{ij}^k - x_{ji}^k \right) - q_i^k \right] \tag{4}$$

The efficiency conditions are

$$v_{ij} \begin{Bmatrix} = \\ \geq \end{Bmatrix} 0 <=> a \cdot \mu_{ij} \begin{Bmatrix} < \\ = \end{Bmatrix} k_{ij} \tag{5}$$

$$x_{ij}^k \begin{Bmatrix} = \\ \geq \end{Bmatrix} 0 <=> \lambda_i^k - \lambda_j^k \begin{Bmatrix} < \\ = \end{Bmatrix} \mu_{ij} \tag{6}$$

Since the $\lambda_j^k$ occur only as differences, they are arbitrary up to an additive constant. Once more we may choose $\lambda_k^k = 0$.

Then $\lambda_i^k$ will represent the cheapest fare from i to k, when $\mu_{ij}$ is the charge for traveling from i to j on link ij.

Condition (6) advises travelers to choose the cheapest – or if $k_{ij}$ and hence $\mu_{ij}$ are in time units – the fastest routes.

Condition (5) states that vehicles be operated only on links where charges collected from travelers cover costs. These charges $\mu_{ij}$ will be positive only where capacities are utilized, thereby forcing the public transportation system to choose frequency of service so as to achieve full loads (or, in practice, reasonable load factors). When the condition is added that vehicles must complete round trips and may not accumulate anywhere

$$0 = \sum_j v_{ij} - v_{ji} \tag{7}$$

a term

$$\sum_i \rho_i \left( \sum_j v_{ji} - v_{ij} \right) \tag{8}$$

is added to the Lagrangean and the efficiency conditions for vehicles are modified to

$$v_{ij} \begin{Bmatrix} = \\ \geq \end{Bmatrix} 0 <=> -k_{ij} + a\mu_{ij} + \rho_j - \rho_i \begin{Bmatrix} < \\ = \end{Bmatrix} 0 \tag{9}$$

Here $\rho_j$ is the economic value of having a vehicle in location j and the value difference $\rho_j - \rho_i$ is added to the fare income in the calculation for recovering costs $k_{ij}$.

These differences will cancel out on any completed round trip.

When travelers choose routes $\tilde{x}_{ij}^{k}$ in advance i.e. independent of the vehicle routing by carriers, the cost minimization problem of the carrier becomes simplified

$$\underset{v_{ij} \geq 0}{\text{Min}} \sum k_{ij} v_{ij}$$

$$a\, v_{ij} \geq \sum_k \tilde{x}_{ij}^{k}$$

$$\sum_j \left(v_{ij} - v_{ji}\right) = 0$$

but total cost is necessarily smaller when the minimization also includes choices by travelers of cheapest routes $x_{ij}^{k}$.

Since operating costs of the public transportation system is exactly recovered from fares on the cheapest routes for travelers, the object of minimizing cost to travelers is the same as that of minimizing cost of the system operation.

N B When trip demands of travelers are balanced in either direction

$$q_i^k = q_k^i$$

then the flows can also be balanced

$$x_{ij} = x_{ji} \qquad\qquad v_{ij} = v_{ji}$$

so that the condition

$$\sum_j \left(v_{ij} - v_{ji}\right) = 0 \tag{7}$$

is automatically satisfied and can be dropped.

## Mathematical Note 13.4
## Capacitated Hub and Spoke Systems

In airline transportation it is now customary to specify some (or just one) transfer locations, as hubs. These permit the collection of flows from many origins to numerous destinations by linking them with hub(s), affording better load factors and enabling greater frequency of flights. The same principle had been used before in other forms of passenger transportation e.g. municipal bus systems. Hubs tend to be crowded and capacity limits must be observed. Let

$$b_i \geq \sum_j \left(v_{ij} + v_{ji}\right) \tag{1}$$

represent an upper bound on incoming plus outgoing vehicle flow in a hub i. Adding these constraints to the Lagrangean (12.3.4)

$$+\sum_i \gamma_i \left[ b_i - \sum_j \left( v_{ij} + v_{ji} \right) \right]$$

will change the efficiency conditions for vehicle routing

$$v_{ij} \begin{Bmatrix} = \\ \geq \end{Bmatrix} 0 <=> a\,\mu_{ij} \begin{Bmatrix} < \\ = \end{Bmatrix} k_{ij} + \gamma_i + \gamma_j$$

The effect of rents $\gamma_j$ being charged to vehicles using hub j is thus to raise the charges $\mu_{ij}$ that make up the final fares $\lambda_i^k$ on cheapest routes from i to k. Rents or user charges $\gamma_i$ for capacitated hubs, may be found e.g. by an auction for landing and departure rights at j, and this is a method of achieving an efficient allocation of airport capacities in air transportation by market methods.

**Mathematical Note 13.5**
**Refraction of Traffic**

It has been remarked first by Palander (1935) and then Stackelberg (1938) and Lösch (1940) that the choice of shortest paths in two-dimensional space by traffic entering a major (one-dimensional) route requires traffic to approach this route at an angle (figure 1).

**Fig. 13.5**

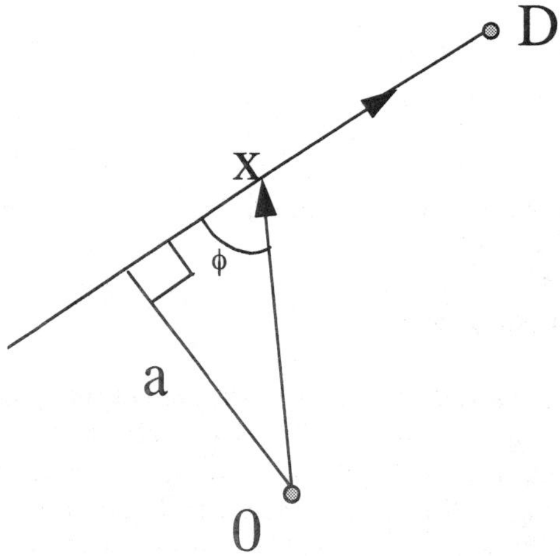


Let traffic originate from 0 at a distance a normal to the super highway which goes directly to destination D. Then $\frac{a}{\sin \phi}$ is the distance traveled outside the highway at costs $k_1$ to save distance a ctg $\phi$ at cost $k_0$ on the highway

$$\underset{\rho}{\mathrm{Min}} \frac{k_1}{\sin\phi} - \frac{k_0}{\mathrm{ctg}\phi}$$

$$0 = -\frac{k_1 \cos\phi}{\sin^2\phi} + \frac{k_0}{\sin^2\phi}$$

yielding

$$\cos\phi = \frac{k_0}{k_1}$$

the law of refraction of traffic, comparable to the refraction of light by Huygens' principle.

## References


Beckmann M (1967) On the Optimal Location of Highway Networks. Quantitative Geography Part I (Garrison WL and Marble DF eds). Evanston, Ill.: Northwestern Univ. Press

Beckmann MJ, McGuire CB, Winsten CB (1956) Studies in the Economics of Transportation. Yale University Press, New Haven

Boyce D, LeBlanc J, Chou Ks (1988) Network Equilibrium Models of Urban Location and Travel Choices: A Retrospective. Journal of Regional Science 28, 2:159–183

Dafermos S (1980) Traffic Equilibrium and Variational Inequalities. Transportation Science 14, 1:42–54

Domencich TA (1975) Urban Travel Demand. A Behavioral Analysis. North Holland, Amsterdam

Ford LR, Fulkerson DR (1962) Flows in Networks. Princeton: Princeton Univ. Press

Herman R (1961 (ed) Theory of Traffic Flow. Elsevier, New York

Koopmans TC (1949) Optimum Utilization of the Transportation System, Econometrica 17, Supplement 136–146

Manheim ML (1979) Fundamentals of Transportation Systems Analysis. MIT Press, Cambridge

Meyer JR, Kain JF, Wohl M (1963) The Urban Transportation Problem. Harvard University Press, Cambridge

Mohring HD (1976) Transportation Economics. Cambridge, Ma.: Ballinger

Mosler KC (1987) Continuous Location of Transportation Networks. Springer-Verlag, New York

Palander T (1936) Beiträge zu Standorttheorie. Alomquist and Wicksell, Uppsalla

Puu T (1979) The Allocation of Road Capital in Two-Dimensional Space. North-Holland, Amsterdam

Stackelberg von H (1938) Das Brechungsgesetz des Verkehrs. Jahrbücher für Nationaloek. und Statistik 148:680–696

Starckie DN (1976) Transportation Planning, Policy and Analysis. Pergamon, New York

Wardrop JG (1952) Some Theoretical Aspects of Road Traffic Research. Proceedings of the Institution of Civil Engineers, Part II:325–378

# CHAPTER 14

# The Spatial Economics of Knowledge

At the heart of location theory have been those economic activities that require the transportation of goods and materials. In the case of service industries the focus is on the transportation of persons instead.

However, of increasing importance for location has become the access to information and knowledge.

The distinction between information and knowledge is loose and fluid. Roughly, information is perishable while knowledge is intended to be durable. The absorption of information is to a first approximation costless, while the digestion of knowledge requires effort, i.e. labor.

Those activities where knowledge is of particular importance are sometimes called the knowledge based or *knowledge industries*. Taking a broad view of the role in economic activities of knowledge we consider in this chapter the spatial economics of knowledge.

We shall study knowledge production and knowledge distribution as economic activities. (Knowledge consumption or rather utilization – since knowledge is never used up – are too universal and non-specific to be distinguished as a separate type of economic activity). First we consider their locations as given and look at interaction and market areas. A distinction must be made between pure basic scientific knowledge, to be considered first, and applied knowledge.

## Knowledge Production

Knowledge production is for the entire world of qualified persons: scientists, scholars, even the curious public, and not for any localized market or market area. When knowledge becomes embodied in scholarly publications, its distribution incurs costs of transportation. The charges for scientific journals are typically of the uniform delivered pricing type. When national boundaries are crossed, there are rarely any customs duties, but booksellers tend to use exchange rates unfavorable to buyers. Also there are often delays in the access to foreign journals. These are but minor impediments to the dissemination of new scientific knowledge. Distance as such should therefore not affect the exchange of scientific ideas. Language does however, and language barriers tend to arise with increasing distance. In many fields English is now becoming the language of international scientific discourse – somewhat to the chagrin of the Francophone world. If citation is used as a measure of such exchange, distance should be irrelevant to citation frequencies. This is not borne out by empirical evidence, however, not even within a country as homogeneous as the United States (Beckmann and Persson, 1996). What could account for this?

Undoubtedly a *personal element* is present in acquaintance with other scientist's work as revealed by citation. Personal acquaintance tends to decrease with distance, as discussed in chapter 1. There may also be an institutional component here. Interaction between institutions is sometimes formalized, say within a given state in the US.

Distance plays an even stronger role in the more intensive types of scientific interaction, the most intensive of which is collaboration resulting in joint publication (cf. math. note 14.2). Once more an analysis of joint publications by authors from Swedish universities reveals distance as a significant factor in a gravity model. The more centrally located scientists engage in more joint work with outside colleagues, since here again "interaction decreases with distance" (Andersson and Persson, 1993).

In applied research there are indeed market relations. Communication or collaboration tend to be not freely chosen but dictated by the demands of the user, who is also a buyer. Rapid communication being of paramount importance, this encourages close proximity between researcher and sponsor, producer and user. The research laboratories of business firms are therefore never far from company headquarters (see below). Aside from rapid access to results, management likes to keep an eye on what their research personnel is doing.

To consider the location of knowledge production – both normative and positive – in detail, distinctions are needed between scientific knowledge and general popular knowledge and again between pure basic knowledge and applied knowledge or technology.

At first sight pure basic research appears to be a footloose industry. Now pure basic knowledge is properly produced in universities. The combination of graduate teaching with research is not accidental but synergetic. Research activity benefits from presentation to an audience of young scholars, and graduate instruction is effectively performed only by those who are themselves actively advancing a field. We are thus brought to studying the optimal location of graduate schools, or at any rate the actual location of successful (= productive) graduate schools. Transportation costs for their students are not a consideration here since the value of the product is orders of magnitude greater than these costs.

Proximity to other graduate schools or research facilities may be a factor to professors, as it affects opportunities for collaboration. Most collaboration occurs however with colleagues in the same university.

Why do not all graduate schools or knowledge producers move together in a single location for maximal interaction?

First, their dependence on financial support – there is no market for basic pure knowledge – means that funding agencies, usually the government will keep them apart for various reasons. Funding by individual states means locating in that state. Even when the national or federal government is the main source of support, regional policy will favor decentralization. Moreover the intention to differentiate its product means that a new graduate school will seek to avoid too close a proximity to other established schools. Finally a university having a college besides its graduate school will seek a location suitable for its college (see below, Knowledge Distribution).

To attract a distinguished faculty, a location in or near a major metropolitan area turns out to be important, for professors tend to be consumers of cultural goods and services, and these are found only in central places of the higher ranks.

While some colleges were founded intentionally in places rather removed from the temptations of the big cities, those that have turned later into research universities are invariably located in or next to those metropolitan areas that can boast of a rich cultural life, to which these universities will in turn contribute on their part.

It has also been found that successful research universities require locations well served by air transportation (Andersson 1988). In a central place system this is true once more for central places of high rank. Periphery must then refer to the lower ranking central places and not necessarily to locations remote in the sense of being at the boundary of a country or region.

A modern type of institution for knowledge production is the "think tank". Here scholars on leave from their home institutions will spend a year or so in productive thought. An isolated location remote from distractions is deliberately sought, and a good climate is desired. Thus California has become home to several think tanks.

Consider now the production of applied knowledge (= applied research) and the development of new technology. These activities are (in the language of location theory) market oriented or more precisely client oriented. In-house research activities, i.e. those performed in research laboratories that are divisions of a business firm, are placed conveniently near company headquarters. GM which at one time owned research facilities in Santa Barbara, California, has since moved all its research to its GM Tech Center in Warren, Michigan, close to company headquarters in Detroit. The famous Bell Laboratories are in Murray Hill, New Jersey, convenient to the headquarters of AT&T. These examples could be multiplied.

Laboratories for applied research that are not affiliated to any particular business firm such as Battelle Institute are market oriented and thus drawn to locations that offer convenient access to actual and potential clients, i.e. major business firms, and are for these reasons also attracted to large metropolitan areas.

Concentration in a small area does occur in knowledge based industries such as computers (Silicon Valley, Route 128) gene technology or the fashion industry (New York's garment district). The aim is maximal access to information and the diffusion of new knowledge.

Consider finally the production of general popular knowledge by authors of books aiming to write not worst selling monographs but best sellers. They are the last of the truly independent economic agents, not tied to any particular organization or location. Their market is the world. Thus they are indeed footloose and free to choose a location that suits their own tastes: one offering cheap cost of living – to a poorer or beginning writer, and a luxurious one with optimal climate and amenities – cultural or otherwise – in the case of the rich and successful author: the French Riviera or certain islands or luxury resorts. (Las Vegas surprisingly can cater to both).

A few remarks may be added on the book publishing industry. Transportation cost of the product dictates a central location for the printing and distribution of books. The principle business of the publisher: acquisition and editing of manuscripts could be considered a footloose activity. But concentration in a large metropolitan area, or rather in a high ranking central place such as New York or Chicago has been the result. It appears that publishing, particularly of fiction, is an activity highly influenced by fashion. Hence access to the latest information is essential and as in the case of other fashion goods it means close proximity to similar producers. While this does not apply to scientific books, publishers often produce both fiction and scientific works in order to reap economies of scope.

The editing of scientific journals on the other hand is truly footloose and often located at a graduate school, where the journal in question was first conceived and launched.

To sum up, since knowledge is not free floating but must be embodied in either persons or documents, activities that require knowledge as an input depend on access to knowledgeable persons or to documents. With modern communication facilities access to documents is available anywhere at practically the same cost. Hence, this cannot influence location. Access to knowledge embodied in persons is critical in the production of new knowledge but important also to other knowledge using economic activities or industries. This is the source of scale economies that are reaped by the institutionalization of academic departments and research laboratories. In quickly advancing fields further productivity gains may be reaped by scientific collaboration with colleagues in other institutions. Since face to face interactions is then still essential, accessibility to and from other knowledge centers is a locational advantage enjoyed by those knowledge producing locations that lie in a (regional, national, continental or world wide) center, and is somewhat but not entirely denied to those in the periphery. Here it should be noted that the areas that may be viewed as central have themselves undergone a process of expansion that no doubt will continue into the future.

## Knowledge Distribution

The *distribution* of new knowledge depends on both oral and written transmission. A distance effect, however weak, is bound to occur even in the distribution of written material (se above).

Oral distribution of knowledge occurs as teaching, seminar presentation to colleagues in the home institution or in inviting institutions, or as papers delivered at conferences, conventions or congresses. Conference sites may be fixed or, more commonly, be rotated to accommodate participants from different locations and with different preferences for meeting sites. Conferences are a market oriented activity since attendance will depend on accessibility (and sometimes amenities).

When teachers were cheap, wealthy families could engage their own tutors. But now it is far more economical to have a teacher address several, sometimes a large number of students, and to assemble these in a fixed location, a school, rather than have the teacher circulate among potential listeners. The so-called Chautauqua circuit still exists for lecturing to adult audiences about popular subjects by well advertised lecturers as a type of summer recreation, but it plays no part in serious scientific education.

Economies of scope encourages the locating together of teaching facilities and teachers of several subjects in one institution, a school, college or university. Whether room and board are also provided, depends on preference and traditions rather more than on economic considerations of efficiency alone.

Consider now the distribution of scientific knowledge at the college level. Unlike graduate school, college attendance depends on proximity to customers. The fact that attendance at a particular college declines with distance was noticed early by students of gravity laws (Carrothers 1966).

For colleges it seems that not only quality and the tuition cost but also transportation cost of students will matter to their clients, i.e. the parents of students. Hence colleges are a dispersed industry, whose market areas will overlap, but with market shares clearly declining with distance.

Schools and colleges are not profit seeking businesses, and thus the Löschian theory of market equilibrium under free entry does not apply. Rather, schools and even institutions of higher learning, are typically state organized and their market areas are influenced by regional policy and thus not entirely dictated by considerations of costs and benefits.

The air fare minimizing solution for the American Economic Association meetings in the US turns out to be Pittsburgh, Pa (Siegfried and Nelson 1979) but has rarely been adopted.

"Summer schools" or similar gatherings for somewhat longer periods tend to be attracted to locations of a recreational (and more remote) character. Table 15 summarizes the locational orientation of knowledge.

**Table 15.** Locational orientation of knowledge activities

| ACTIVITY | ORIENTATION |
|---|---|
| 1. Research university | cultural centers (by personnel), hence large metropolis' |
| 2. Knowledge using firms | Research universities preferably with Nobel laureates |
| 3.1 Research laboratories: corporate divisions | Corporate headquarters |
| 3.2 Independent research laboratories | Clients, hence business centers |
| 4. Think tanks | Isolation in naturally attractive setting |
| 5. Congresses | Central places with air connections and hotel space |
| 6. Colleges | Population |
| 7. Summer schools | Small college campuses |
| 8. Publishers | Other publishers, hence large metropolis' |
| 9. Writers | Footlose<br>Rich: Fashionable riviera type resorts<br>Poor: Low cost of living |
| 10. Retired Scholars | Footlose: Universities, retirements havens |

## Mathematical Note 14.1
## Knowledge Diffusion as a Spatial Process

Since knowledge comes in many forms, the processes of knowledge transmission must also be studied for various contexts. One instance that has attracted much attention by both econometricians (Griliches, 1957) and geographers (Hägerstrand, 1967) is the spread of new agricultural techniques through rural areas. In this process, adoption seems to follow mainly upon observation of success at the farms of neighbors. This mechanism is mathematically similar to the spread of an epidemic through contact between an infected with an uninfected person.

Two modes of knowledge need to be distinguished, that embodied in documents, i.e., publications and that knowledge which is still embodied only in persons. The prestige motivated activity of scholars and scientists in scientific research demands that ultimately all results, i.e., scientific knowledge be published.

The distribution of documents is not a spatial process, for space can here in a first approach, be disregarded. Subscription costs for periodicals do not depend much on distance, nor is there much difference in the time required for scientific journals to reach even remote places. Books are sometimes subject to customs duties. But what is essential is the size and budget of a library where access is sought, not its location.

On the other hand, publication of new knowledge may be tardy and incomplete. Scientific progress, particularly in rapidly advancing field, depends very much on interpersonal communication.

Access to knowledge in persons is much more selective than that to knowledge in publications. While the latter requires no more than the availability of a research library of sufficient size (plus access to the intenet): access to persons tends to be restricted to a network of interpersonal relations based largely but not exclusively, on personal acquaintance. The circulation of prepublications, i.e., documents such as discussion papers falls somewhere between that of personal knowledge and formally published knowledge.

The term "knowledge networks" can be used to refer to the set of links by which either these papers or personal communications are transmitted. The nodes of a knowledge network are termed "knowledge centers." In principle any institution which produces or distributes knowledge should be considered a knowledge center, and in fact every such institution is connected by some knowledge exchange with some other knowledge center and is therefore a knowledge center in the first sense of being a node of a knowledge network.

That knowledge is produced not by individuals in isolation but rather in organized institutions such as university departments or research laboratories is essentially due to economies of scale. One factor contributing to scale economies was mentioned already: the need for access to a research library. To this should be added the stimulation and critical comment afforded by the presence of colleagues engaged in research on the same or related problems. In rare cases a single eminent scholar working without assistance or technical support staff in theoretical research, e.g., mathematics – may constitute a single person knowledge center.

Our object is to model the diffusion of knowledge through a knowledge network as a spatial process. We wish to determine $P_i(t)$ the probability that a particular piece of knowledge is known and accepted in knowledge center i by time t. The contacts of i to other knowledge centers j through the network will be described by parameters $h_{ji}$ denoting the intensity of interaction between j and i as follows. Knowledge is available in i at time $t+\Delta t$ either because it was there already by time t or because it was transmitted from j in time $\Delta t$

$$P_i(t+\Delta t) = P_i(t) + [1 - P_i(t)]P_j(t)h_{ji}\Delta t \tag{1}$$

Letting $\Delta t \to 0$ this becomes

$$\frac{dP_i}{dt} = P_j h_{ji} \left[1 - P_i\right] \tag{2}$$

When several knowledge centers j connected to i are considered, one has instead

$$P_i(t + \Delta t) = P_i(t) + \left[1 - P_i(t)\right] \sum_j P_j(t) h_{ji} \Delta t + 0(\Delta t) \tag{3}$$

where the summation extends over all j linked to i in the network,

$$\frac{dP_i}{dt} = \left[1 - P_i\right] \sum_j h_{ji} P_j \tag{4}$$

Integrating (2)

$$\frac{1}{1 - P_i} \frac{dP_i}{dt} = h_{ji} P_j$$

$$-\log(1 - P_i)\Big|_0^t = h_{ji} \int_0^t P_j dt$$

Assume that the knowledge considered is not known in i at t = 0

$$P_i(0) = 0 \qquad \log(1 - P_i(0)) = 0$$

$$1 - P_i = e^{-h_{ji} \int_0^t P_j dt} \tag{5}$$

Assume, moreover, that this knowledge was known in j already at time t = 0, then

$$P_i(t) = 1 - e^{-h_{ji} t} \tag{6}$$

The expected time $\bar{t}_i$ for knowledge to be known and accepted in i is then

$$\bar{t}_i = \int_0^\infty t \frac{dP_i}{dt} dt = \int_0^\infty (1 - P_i) dt = \int_0^\infty e^{-h_{ji} t} dt = \frac{1}{h_{ji}} \tag{7}$$

The expected transmission time is thus inversely proportional to the intensity of contact $h_{ji}$. The corresponding results for transmission through alternative links ji to knowledge center with $p_j = 1$ are

$$P_i(t) = 1 - e^{-\sum_j h_{ji} t} \tag{8}$$

$$\bar{t}_i = \frac{1}{\sum_j h_{ji}} \tag{9}$$

Formulas (7) and (9) show the importance of linkages $h_{ji}$ in a knowledge network. It is conceivable that all nodes in a finite network have the same degree d(i), i.e., the same number of direct links to other nodes and that the intensities $h_{ji}$ are all equal = h.

In that case an uneven initial distribution of knowledge will soon be followed by an equal distribution of knowledge among all knowledge centers.

A more likely situation is, however, that nodes i will differ in regard to both the number and the intensity of links $h_{ji}$ to other nodes j so that a distinction between periphery and center will emerge. This should not be interpreted in a narrow geographical sense. In the US the many smaller universities of the Middle West (apart from giants such as the University of Chicago and the Big Ten) are in fact in the periphery of the knowledge network while the research universities of the East and West Coast are properly in the center. Similarly Sweden has more scientific links with the US and Central Europe than have some of the countries in the very center of the European Continent.

### Mathematical Note 14.2
### Collaboration and Distance

In scientific collaboration two agents (scientists) supply labor inputs y, z measured in units of time into the production of a joint output q

$$q = f(y,z)$$

The production function is considered to have the usual properties

- positive marginal products

  $f_y > 0 \qquad f_z > 0$

- diminishing returns to substitution

  $\begin{pmatrix} f_{yy} & f_{yz} \\ f_{zy} & f_{zz} \end{pmatrix}$ is negative definite. In particular

  $f_{yy} < 0 \qquad f_{zz} < 0$

- constant returns to scale

  $f(\lambda y, \lambda z) = \lambda f(y,z)$

Let u, v represent the opportunity cost of time to the agents and let t be the time required to realize one hour of actual input. In the case of collaboration between agents in different locations separated by distance r, we postulate

$$t = kr \tag{1}$$

where the proportionality factor k reflect the speed and ease (or rather their inverses) of transportation and/or communication.

The opportunity cost u,v are measured in the same units as the output q. In the case of pure basic research their dimension is presumably "scientific prestige", in the case of applied research it should be money.

Rational behavior of the agents means to the first agent

$$\max_{y \geq 0} f(y,z) - uty \tag{2}$$

and to the second agent

$$\max_{z \geq 0} f(y,z) - vtz \tag{3}$$

(Here it is assumed that the scientific gain of prestige of the researcher goes to both agents equally). For convenience let f be Cobb Douglas

$$f(y,z) = by^{\beta}z^{\alpha} \qquad \alpha + \beta = 1 \tag{4}$$

From (2) using (4)

$$\beta b\left(\frac{z}{y}\right)^{\alpha} = ut \tag{5}$$

and from (3)

$$\alpha b\left(\frac{y}{z}\right)^{\beta} = vt \tag{6}$$

Dividing (6) into (5)

$$\frac{\beta}{\alpha}\frac{z}{y} = \frac{u}{v}$$

or

$$z = \frac{\alpha}{\beta}\frac{u}{v} \cdot y$$

Substituting into the objective function (2) of the first partner, the yield from input y is seen to be

$$\left[b\left(\frac{\alpha}{\beta}\frac{u}{v}\right)^{\alpha} - tu\right]y \tag{7}$$

$$= \left[b\left(\frac{\alpha}{\beta}\frac{u}{v}\right)^{\alpha} - kru\right]y \tag{7a}$$

$$= \left[bh - kru\right]y \qquad \text{(say)} \tag{7b}$$

It decreases linearly with distance.

Now let b, the productivity factor of the collaboration, be a normally distributed random variable and a an acceptance level. The probability of a collaboration being advantageous to the first agent is then

$$\text{pr}(hb - ukr \geq a) = \text{pr}\left(b \geq \frac{a}{h} + \frac{uk}{h} r\right)$$

$$= 1 - N\left[\frac{1}{\sigma}\left(\frac{a}{h} + \frac{uk}{h} r - \mu\right)\right] \tag{8}$$

in terms of the standard normal distribution. With a logistic approximation

$$1 - N(x) \doteq \frac{1}{1 + e^{mx}} \tag{9}$$

$$m = \frac{2\sqrt{2}}{\pi} \tag{9a}$$

The probability of a collaboration at distance r is thus approximately

$$\frac{1}{1 + e^{\frac{m}{\sigma}\left[\frac{a}{h} - \mu + \frac{uk}{h} r\right]}}$$

$$\doteq ce^{-\gamma r} \tag{10}$$

showing an exponential distance effect.

When this probability is multiplied by the number of scientists (in a given field) at either location, the result is once more a gravity formula of type (1.2.2.)

## References


Andersson Å (1988) Universitet. Stockholm Institute for Futures Studies

Andersson Å, Batten D, Kobayashi K (1988) Knowledge, Nodes and Netw.rks, Working Paper CERUM, Umeå, Sweden

Andersson Å, Persson O (1993) Networking Scientists. Annals of Regional Science 27:11–21

Beckmann M (1994) On Knowledge Networks in Science: Collaboration Among Equals. Annals of Regional Science 28:233–242

Beckmann M, Persson O (1996) A Note on Citation and Distance. Institute for Futures Studies, Stockholm

Beckmann M, Johanson B, Snickars F, Thord R (1998) Knowledge and Networks in a Dynamic Economy. Springer-Verlag, Heidelberg

Brown LA (1968) Diffusion Processes and Location. Bibliography Series 4. Regional Science Research Institute, Philadelphia

Griliches Z (1957) Hybrid Corn: an Exploration in the Economics of a Technical Chance. Econometrica 25:501–522

Hägerstrand T (1967) Innovation Diffusion as a Spatial Process. University of Chicago Press, Chicago

Katz JS (1994) Geographical Proximity and Scientific Collaboration. Scientometrics, Vol 31, No 1:31–43

Luukkonen T, Persson O, Sivertsen G (1992) Understanding Patterns of International Scientific Collaboration. Science, Technology and Human Values, Vol 17, No 1, Sage Publications Inc.

Persson O (1991) Regional Collaboration in Science. Data on Nordic Coauthorships 1988–1990. CERUM Working Paper CWP-1991:16

Siegfried JJ, Nelson L (1979) Minimizing AEA Convention Transportation Cut. AER 69,5:938–988

Zucker LG, Darby MR (1998) Enterpreneurs, Star Scientists, and Biotechnology. NBER Reporter, Fall, 7–10

# Welfare

So far we have studied economic behavior in a spatial setting motivated by profit seeking. In this chapter we reconsider spatial economics from the viewpoint of welfare maximization.

Since perfect competition is the exception rather than the rule in spatial markets, the well-known theorems on Pareto optimality in general competitive equilibrium do not apply, and each case must be examined on its own merits. The examples of market failure to be examined below show a case for economic policy. Policy is needed already in deciding such technical matters as which side of the streets to drive on. The case for road user charges, to name but one example, is less well understood by the public. This need not mean more taxation but could be an optimal source of revenue for road financing (note 15.9). The main role of road pricing is however that of an efficient device for allocating scarce road capacity in the face of congestion. The welfare loss from traffic delays could then be greatly reduced and the utilization of underused public transportation systems be effectively improved.

One key question about spatial markets is whether too many or too few firms exist in a dispersed industry, i.e. whether there is an oversupply or a dearth of services, and goods offered.

The (non-spatial) theory of monopolistic competition suggests an oversupply. This is confirmed for markets with mill pricing or marginal cost pricing, but rejected in the case of uniform pricing. The equilibrium market size is the same under mill and uniform pricing (for linear demand: since profits depend identically on the market radius R), but the welfare maximizing market size is smaller for uniform pricing, (see note 15.5) and so is welfare per area or household (cf. 15.7) since transportation cost is wasted by uniform pricing.

Welfare optimizing pricing should, of course, be marginal cost pricing. But that does not allow for the recovery of fixed costs, when costs have the simple linear structure that is conventionally assumed in location theory. A second best solution to that is a suitable combination of marginal cost and perfectly discriminatory pricing (note 15.6).

When comparing conventional pricing systems, it is no surprise that mill pricing creates more welfare than uniform pricing for the same quantities sold, since uniform pricing encourages waste of transportation. Although in a market of given size, mill pricing generates more welfare than discriminatory pricing, total welfare generated under discriminatory pricing is larger because of its larger market size (Holahan 1975) (mathematical note 15.7).

The role of policy in preventing uneconomical city size is less clear. In practice, housing schemes and subsidies to transportation systems would seem to encourage uneconomically large city sizes. As note 15.8 suggests, a wage structure reflecting the higher costs of housing and transportation in larger cities could act as a counter force to the uneconomical attraction of larger cities for industries. But the issues of migration and individual choices of residence with all their complexities of quality of life (services, environmental issues) have not been addressed here.

Economic efficiency apart, there is also an aesthetic side to the uses of space that does not make an appearance in the conventional welfare functions of economic theory. It seems that not even the invisible hand would guide decision makers in the direction of creating more beauty. Nevertheless, corrective policies – perhaps in the form of special awards – are conceivable that could improve the aesthetics of the spatial economy.

### Mathematical Note 15.1
### The Optimal Size of a Market

Since total welfare is welfare per spatial market time the number of markets it is actually welfare per unit area that should be maximized (for the same reasons that maximizing profits per area was optimal in a multiple outlet monopoly or collusive oligopoly, see chapter 8).

In spite of well-known objections to the consumers' surplus as a welfare measure (Hicks 1956) it will be appropriate to the partial equilibrium models of location theory to use the consumers' surplus, particularly since location theory works with given demand functions (linear or rectangular) rather than utility functions directly.

The consumers' surplus is related to the demand function

$$q = q(p) \tag{1}$$

as follows. Let $p = p(q)$ be the inverse of (1).

Then the consumers' surplus considered as a function of market price p is

$$u(p) = \int_0^{q(p)} [p(x) - p]dx \tag{2}$$

Thus in the linear case

$$q = a - p \qquad p = a - q$$

$$u(p) = \int_0^{a-p} [a - x - p]dx$$

$$= (a-p)^2 - \frac{1}{2}(a-p)^2 = \frac{1}{2}(a-p)^2 \tag{3}$$

that is the area of the triangle between the horizontal p and the straight line demand curve.

For rectangular demand

$$u(p) = \int_0^1 [a - p] dx = a - p \quad \text{for } p \leq a \tag{4}$$

To determine the optimal market areas, a pricing system p(r) must be specified. We consider first a single market and an arbitrary but given price system p(r). Now welfare is the sum of consumers' surplus and profits

$$W = \int_0^R \left(u(p(r)) + [p(r) - c - k(r)] q(p(r))\right) \rho(r) dr - F$$

The condition for the optimal market radius R in a single spatial market is

$$0 = \frac{\partial W}{\partial R} = u(p(R)) + [p(R) - c - k(R)] q(p(R)) \tag{5}$$

Consider first a system of marginal cost pricing

$$p(r) = c + k\,(r) \tag{6}$$

Then the profit term in (5) disappears and the consumers' surplus u(p) must be zero at the market radius R.

Let a be the price at which quantity demanded reaches zero

$$q(a) = 0 \tag{7}$$

Now

$$0 = u\,(p) = \int_0^{q(p)} [p(x) - p] dx$$

implies that q (p) = 0 since the integrand is positive for quantities $x < q(p)$.

Hence

$$q(p(R)) = 0$$

or

$$p(R) = c + kR = a \tag{8}$$

The market should be extended to the point where quantity demanded at the marginal cost price has fallen to zero.

Consider next mill pricing

$$= \int_0^R \left[u(p_M + k(r)) + (p_M - c) q(p_M + k(r))\right] \rho(r) dr - F$$

$$0 = \frac{\partial W}{\partial R} = \left[u(p_M + k(R)) + (p_M - c) q(p_M + k(r))\right] \rho(R) \tag{9}$$

implies once more that $q(p(R)) = 0$ when $\rho(R) > 0$, or

$$p_M + k(R) = a \tag{10}$$

Although here, too, quantity demanded must reach zero at the market boundary, the market radius for (10) is shorter since

$$p_M - c > 0$$

the profit margin will be positive.

In uniform pricing

$$p(r) = p_U$$

welfare is

$$W = \int_0^R \left(u(p_U) + [p_U - c - k(r)]q(p_U)\right)\rho(r)dr - F$$

and

$$\frac{\partial W}{\partial R} = \left(u(p_U) + [p_U - c - k(R)]\right)q(p_U)\right) \cdot \rho(R) > 0 \qquad \text{at } c + k(R) = p_U \tag{11}$$

This means that welfare can still be increased by expanding the market beyond the point where marginal cost to the firm just equals the price $p_U$. In fact for linear demand

$$q = a - p$$

the welfare maximizing market radius is given by

$$0 = \frac{1}{2}(a - p_U)^2 + (p_U - c - kR)(a - p_U)$$

or

$$R = \frac{1}{k}\left(\frac{a + p_U}{2} - c\right)$$

$$> \frac{1}{k}(p_u < a - c)$$

for profit maximizing uniform and mill prices

*Welfare per area*

Welfare per area is

$$\frac{W}{A} = \frac{\int \left(u(p(r)) + [p(r) - c - k(r)]q(p(r))\right)\rho(r)\,dr - F}{\int \rho(r)dr} \tag{12}$$

and maximization with respect to R requires

$$0 = \left(u(p(R)) + [p(R) - c - k(R)]q(R)\rho(r)\right)A(R)$$

$$- \rho(R)\left(\int \left(u + [p - c - k(r)]q\right)\rho dr - F\right) \tag{13}$$

With positive population density

$$\rho(R) > 0$$

this simplifies to

$$u(p(R)) + [p(R) - c - k(R)]q(p(R)) = \frac{W}{A} \tag{14}$$

Welfare per unit area at the market boundary (without fixed cost) must equal the average welfare in the market area (including fixed cost). This condition that marginal anything be equal to average anything is a well-known one for maximizing the average of anything.

Comparing (14) and (5) reveals that welfare per area is maximized for smaller areas.

## Mathematical Note 15.2
## Welfare Maximizing Market Size for Rectangular Demand, Independent of Price

It was shown in chapter 3 equation (3.4.3) that for given prices the profit maximizing market radius is that at which

$$f(p(R)) \cdot [p(R) - c - kR]\rho(R) = 0$$

so that either population has disappeared $\rho(R) = 0$ or marginal cost c+kR has reached the level of prices p(R) or quantity demanded at price p(R) has fallen to zero. The last condition determines the market under mill pricing, marginal cost equals price under uniform and perfectly discriminating pricing. Welfare for a single market area is maximized when similarly, the welfare contribution at the market boundary drops to zero.

$$u(p(R)) + f(p(R))[p(R) - c - kR] = 0$$

Here u(p) is the indirect utility function, i.e. the achieved utility as a function of price.

When a system of market areas is involved, it is not welfare generated in a particular market area that is relevant, but this times the number of market areas or equivalently welfare per area or per household in the area that should be maximized.

Now welfare goes both to consumers, as a utility or consumers' surplus (see 15.1) and to firms as profits. They both depend on prices except in the interesting case of rectangular demand to be considered in this note. For rectangular demand the consumers' surplus is simply the difference between reservations price a and actual price p(r). For any pricing system p(r)

$$\frac{W}{M} = \frac{\int_0^R \left([a - p(r)] + [p(r) - c - kr]\right)\rho(r)dr - F}{\int_0^R \rho(r)dr} \tag{1}$$

where the first term under the integral of the numerator is consumers' welfare and the second term the firms' gross profits.

Since p(r) drops out

$$\frac{W}{A} = \frac{\int_0^R [a - c - kr]\rho dr - F}{\int \rho dr} \tag{2}$$

$$\operatorname{Max} \frac{W}{A} = a - c - \underset{R}{\operatorname{Min}} \frac{k\int r\rho(r)dr + F}{\int \rho(r)dr}$$

Welfare maximization has thus been reduced to the minimization of average cost per household served

$$= \underset{R}{\operatorname{Min}} \left[ k\bar{r}(R) + \frac{F}{M(R)} \right] \tag{3}$$

where

$$\bar{r}(R) = \frac{\int r\rho(r)dr}{\int \rho(r)dr}$$

is average distance and

$$M(R) = \int \rho(r)dr$$

is population served.

For circular market areas this means

$$\operatorname{Min}\ k\frac{2}{3}R + \frac{F}{\pi\rho R^2}$$

$$0 = \frac{2}{3}k - \frac{2F}{\pi\rho R^3}$$

$$R = \sqrt[3]{\frac{3}{\pi}\ \frac{F}{k\rho}}$$

or with standardization k = 1

$$R_0 = .9847\sqrt{\frac{F}{\rho}} \tag{4}$$

For hexagonal market areas (cf. table 2.2)

$$\operatorname{Min}\quad .702\,R + \frac{F}{2\sqrt{3}\rho R^2}$$

$$0 = .702 - \frac{F}{\sqrt{3}\rho R^3}$$

$$R_H = \sqrt[3]{\frac{1}{.702 \cdot \sqrt{3}} \frac{F}{\rho}} = .9369 \sqrt[3]{\frac{F}{\rho}} \tag{5}$$

With Manhattan metric

$$\text{Min} \frac{2}{3} kR + \frac{F}{2\rho R^2}$$

$$0 = \frac{2}{3} k - \frac{F}{\rho R^3}$$

$$R_{\#} = \sqrt[3]{\frac{3}{2} \frac{F}{\rho}} = 1.1447 \sqrt[3]{\frac{F}{\rho}} > R_0 \tag{6}$$

These should be compared to the market size realized e.g. in Löschian imperfect competition with free entry. For rectangular demand the monopoly or price leader's mill price is

$$p_M = a - kR$$

so that profits are

$$G_M = k\left[\frac{a-c}{k} - R\right]M - F$$

or standardized

$$= (1 - R)M - F \tag{7}$$

For circular markets

$$G_0 = (1 - R)\pi\rho R^2 - F$$

yielding the equilibrium condition

$$\frac{F}{\pi\rho} = R^2 - R^3 = \phi_1(R) \tag{8}$$

For hexagonal markets

$$G_H = (1 - R)2\sqrt{3}\rho R^2 - F$$

$$\frac{F}{2\sqrt{3}\rho} = R^2 - R^3 = \phi_2(R) \tag{9}$$

and for Manhattan metric

$$G_{\#} = (1-R)2\rho R^2 - F$$

$$\frac{F}{2\rho} = R^2 - R^3 = \phi_3(R) \tag{10}$$

Rewriting the conditions for welfare optimality

$$\frac{F}{\pi\rho} = \frac{R^3}{3} = \psi_1(R) \tag{4a}$$

$$\frac{F}{2\sqrt{3}\rho} = .35R^3 = \psi_2(R) \tag{5a}$$

$$\frac{F}{2\rho} = \frac{R^3}{3} = \psi_3(R) \tag{6a}$$

and consider

$$\frac{\phi_i - \psi_i}{R^2}$$

$$= 1 - R - \frac{R}{3} = 1 - \frac{4}{3}R > 0 \quad \text{for} \quad R < \frac{3}{4} \tag{11}$$

$$= 1 - R - .35R > 0 \quad \text{for} \quad R < \frac{1}{1.35} = .74 \tag{12}$$

$$= 1 - R - \frac{R}{3} > 0 \quad \text{for} \quad R < \frac{3}{4} \tag{13}$$

Now $R = \frac{3}{4}$ is the profit maximizing R for circles and Manhattan circles (tilted squares), and for hexagonal markets this is .71. Thus, in the relevant range for (11), for (13) always and for (12) almost always

$$\phi > \psi$$

**Fig. 15.2.** Comparing market radius $\hat{R}$ under free entry and the welfare optimizing market radius $R_{opt}$

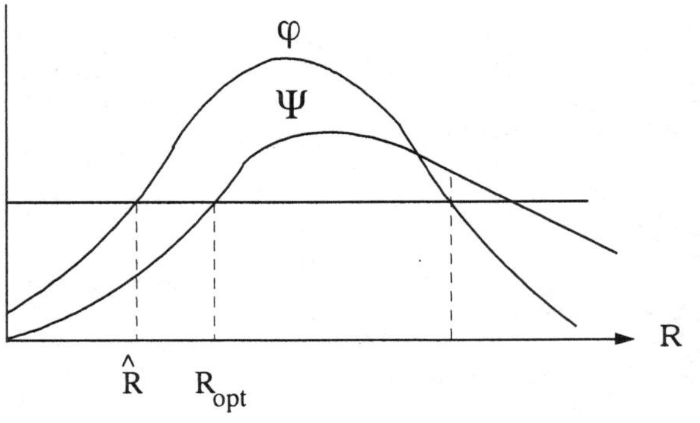

implying (figure 15.2) that the market radii under free entry are smaller than the welfare optimal size, in conformity with the proposition that monopolistic competition with free entry encourages too many firms.

### Mathematical Note 15.3
### Optimal Market Size under Marginal Cost Pricing

In this note we optimize both with respect to pricing and to market size, and show how this may be approximated. Observe first that the proper objective is not welfare in one market area but this times the number of market areas which as noted before becomes welfare per area or welfare per household, the latter being the right criterion even when population densities vary. Thus we should

$$\underset{R}{\text{Max}}\frac{1}{M(R)}\left\{\int_0^R \big(u(p(r))+q(p(r))[p(r)-c-kr]\big)\rho(r)dr - F\right\}$$

The optimal R is then determined by

$$0=\big(u(p(R))+q(p(R))[p(R)-c-dR]\big)\rho(R)\cdot M(R)$$

$$-\rho(R)\cdot\left\{\int_0^R u(p(r))+q(p)(r))[p(r)-c-kr]\rho(r)dr - F\right\}$$

or

$$u(p(R))+q(p(R))[p(R)-c-kr]=\frac{W}{M}$$

$$\text{if } \rho(R)>0.$$

Marginal welfare at the marginal locations R must equal average welfare. Observe that marginal welfare must now be positive and not zero as when maximizing for a single market area. This implies smaller market areas. For constant population density and circular markets the optimal market size under marginal cost pricing may now be calculated for linear demand.

$$\underset{R}{\text{Max}}\frac{1}{\pi\rho R^2}\left[\int \frac{1}{2}[a-c\text{-}kr]^2\, 2\pi\rho r dr - F\right]$$

Note that profits (or losses) are simply – F since with linear cost functions marginal cost equals average variable cost.

Standardizing

$$\underset{R}{\text{Max}}\frac{1}{R^2}\int_0^R (1-r)^2 r dr - \frac{F}{\pi\rho R^2}$$

$$=\underset{R}{\text{Max}}\frac{1}{R^2}\left(\frac{R^2}{2}-\frac{2}{3}R^3+\frac{R^4}{4}\right)-\frac{F}{\pi\rho R^2}$$

Maximizing

$$0 = -\frac{2}{3} + \frac{1}{2}R + \frac{2F}{\pi\rho R^3}$$

giving rise to a quartic equation

$$\frac{2}{\pi}\frac{F}{\rho} = \frac{2}{3}R^3 - \frac{1}{2}R^4 \tag{1}$$

Suppose that on condition of charging marginal costs c to consumers as the mill price, firms received a subsidy x per commodity sold and y F for fixed cost. Can these be chosen so as to achieve the welfare optimum?

At market size R profit s of a representative firm are

$$G(R) = x \int_0^R [a - c - kr]\rho(r)dr - (1 - y)F$$

Standardizing

$$G = x\int_0^R (1 - r)2\pi\rho r dr - (1 - y)F$$

and the equilibrium radius R where profits have vanished is determined by

$$0 = x\left[R^2 - \frac{2}{3}R^3\right]\pi\rho - (1 - y)F \tag{2}$$

To compare with (1) rewrite (2)

$$\frac{2F}{\pi\rho} = \frac{2x}{1 - y}\left[R^2 - \frac{2}{3}R^3\right] \tag{2a}$$

This becomes

$$\frac{2F}{\pi\rho} = \frac{2}{3}R^3 - \frac{4}{9}R^4 \tag{3}$$

or approximately (1) when the subsidies are chosen as

$$\frac{2x}{1 - y} = \frac{2}{3}R \quad x = \frac{R}{3} \quad y = 0 \tag{4}$$

To do this the optimal R must be calculated by the subsidizer first. Notice that a subsidy x per unit sold is necessary but that the subsidy to fixed cost is superfluous.

**Mathematical Note 15.4**
**Optimal Market Size for Linear Demand and Mill Pricing**

The firm's profits in a market (of any shape and metric) of inner radius R – the half distance to its neighbors – was found in chapter 3 equation (3.3.19) and chapter 7 equation (7.1.1) to be

$$G_M = \frac{k^2}{4}\left[\frac{a-c}{k} - \bar{r}(R)\right]^2 M(R) - F \tag{1}$$

Upon standardization

$$G_M = \frac{1}{4}[1-\bar{r}(R)]^2 M(R) - F \tag{1a}$$

For constant population density

$$\rho(r) = 2\pi r$$

$$\bar{r} = \frac{2}{3}R \tag{2}$$

$$M = \pi\rho R^2 \tag{3}$$

$$G_M = \frac{\pi\rho}{4}\left[1-\frac{2}{3}R\right]^2 R^2 - F \tag{4}$$

and the equilibrium condition is

$$\frac{4F}{\pi\rho} = \left(1-\frac{2}{3}R\right)^2 R^2 = R^2 - \frac{4}{3}R^3 + \frac{4}{9}R^4 = \phi(R), \text{say} \tag{5}$$

Consider next the welfare maximal market radius for the same pricing system

$$\underset{R}{\text{Max}} \frac{1}{M(R)}\left\{\int_0^R \left(\frac{1}{2}\left[a - \frac{a+c-k\bar{r}}{2} - kr\right]^2\right.\right.$$

$$\left.\left. + \left[a - \frac{a+c-k\bar{r}}{2} - kr\right]\left[\frac{a+c-k\bar{r}}{2} - c\right]\right)\rho dr - F\right\}$$

Standardizing and assuming constant population density

$$\frac{1}{\pi\rho R^2}\left\{\int_0^R \left[\frac{1}{2}\frac{(1+\bar{r}-2r)^2}{4} + \frac{(1-\bar{r})(1+\bar{r}-2r)}{4}\right] 2\pi\rho r dr - F\right\}$$

$$= \frac{1}{4R^2}\int_0^R (1+\bar{r}-2r)^2 + 2(1-\bar{r})(1+\bar{r}-2r) r dr - \frac{F}{\pi\rho R^2}$$

Completing the square

$$= \frac{1}{4R^2}\int\left(\left[1+\bar{r}-2r+1-\bar{r}\right]^2-\left[1-\bar{r}\right]^2\right)rdr-\frac{F}{\pi\rho R^2}$$

$$= \frac{1}{R^2}\int(1-r)^2 rdr - \frac{1}{4}(1-\bar{r})^2 rdp - \frac{F}{\pi\rho R^2}$$

$$= \frac{1}{2}-\frac{2}{3}R+\frac{R^2}{4}-\frac{1}{8}+\frac{R}{6}-\frac{R^2}{18}-\frac{F}{\pi\rho R^2}$$

$$= \frac{3}{8}-\frac{1}{2}R+\frac{7}{36}R^2-\frac{F}{\pi\rho R^2}$$

Maximizing with respect to R

$$0 = -\frac{1}{2}+\frac{7}{18}R+\frac{2F}{\pi\rho R^3}$$

or

$$\frac{4F}{\pi\rho} = R^3-\frac{7}{9}R^4 = \psi(R) \qquad \text{(say).}$$

Now

$$\phi-\psi = R^2-\frac{4}{3}R^3+\frac{4}{9}R^4-R^3+\frac{7}{9}R^4$$

$$\frac{\phi-\psi}{R^2} = 1-\frac{7}{3}R+\frac{11}{9}R^2$$

$$= \frac{11}{9}\left[R^2-\frac{21}{11}R+\frac{9}{11}\right] > 0 \quad \text{for} \quad R < .65$$

$$= \frac{11}{9}\left[\left(R-\frac{21}{22}\right)^2-\frac{5}{44}\right] > 0 \quad \text{for} \quad R < .65$$

From figure 11.2 it may be seen that this implies a smaller market size under free entry than is welfare optimal, when market size falls below the critical level .65 (less than the maximal .75).

A straight forward, but tedious calculation along the same lines shows that similar results are true also for hexagonal markets and under Manhattan geometry. For rectangular demand functions it was proved already in note 15.2.

## Mathematical Note 15.5
## Welfare Optimal Market Size for Uniform Pricing

Consider the scenario of price taking with uniform prices. Clearly, overlapping market areas cannot be optimal since this wastes fixed cost and, in a random assignment of customers to firms, also transportation cost. It was seen in mathematical note 4.2 that any increase in the density of firms tends to raise prices towards monopoly levels. Therefore, for efficiency markets should be separated and adjacent. For simplicity we consider them to be circular although under free entry they should eventually become hexagonal. Through Bertrand oligopolistic competition, marginal cost is just covered by prices at the boundary, that is at the minimal level consistent with the given market size.

$$p_U = c + kR$$

With circular market areas and constant population density

$$\rho(r) = 2\pi\rho r$$

$$\frac{W}{M} = u(c+kR) + \frac{1}{M}\left[ f(c+kR)k\int_0^R (R-r)\rho(r)dr - F \right]$$

For linear demand functions this is

$$\frac{W}{M} = \frac{1}{2}(a-c-kR)^2 + \frac{1}{\rho\pi R^2}\left[ (a-c-kR)k\int_0^R (R-r)2\pi\rho r dr - F \right]$$

Standardizing

$$= \frac{1}{2}(1-R)^2 + \frac{1}{\pi\rho R^2}\left[ (1-R)\frac{\pi\rho}{3}R^3 - F \right]$$

$$= \frac{1}{2}(1-R)^2 + \frac{1}{3}(1-R)R - \frac{F}{\pi\rho R^2}$$

Maximizing with respect to R

$$0 = -(1-R) + \frac{1}{3} - \frac{2}{3}R + \frac{2F}{\pi\rho R^3}$$

or

$$\frac{2}{\pi}\frac{F}{\rho} = (1-R)R^3 - \frac{R^3}{3} + \frac{2}{3}R^4$$

$$= \frac{2}{3}R^3 - \frac{1}{3}R^4 \tag{1}$$

Under free entry, market equilibrium is given by

$$F = f\left(c + kR\right)\int_0^R k\left[R - r\right]\rho dr$$

or, standardized

$$F = (1 - R)\pi\rho\frac{R^3}{3}$$

$$\frac{3F}{\pi\rho} = R^3 - R^4 \qquad = \phi(R)\ , \qquad \text{(say)} \qquad (2)$$

Rewrite (1)

$$\frac{3}{\pi}\frac{F}{\rho} = R^3 - \frac{1}{2}R^4 = \psi(R) \qquad \text{(say)}$$

and consider

$$\phi - \psi = -\frac{1}{2}R^4 < 0$$

Therefore (cf fig. 15.2 with $\phi$ now below $\psi$) the equilibrium radius is larger than the socially optimal one for uniform pricing at minimum prices for their market size.

We conclude that with uniform pricing too few firms survive the market process under free entry. The smaller market radius for welfare optimization would then not be sufficient to recover fixed cost. Hence a subsidy would be needed. It can not be a lump sum, since that would reduce fixed cost, while to induce a smaller market size under free entry, a larger fixed cost is required. It can be achieved only by subsidizing unit production cost.

**Mathematical Note 15.6**
**Welfare Optimal Pricing**

As before, the welfare measure for consumers is as an indirect utility function, i.e. the consumers' surplus.

To find the welfare optimal price system p(r) in a market of radius R consider

$$\underset{p(r)}{\text{Max}}\int_0^R\left[\int_0^{q(p(r))}(p(y) - p(r))dy + \left(p(r) - c - kr\right)q(p(r))\right]\rho(r)dr - F$$

Differentiation under the integral yields

$$0 = 0 - q'\cdot\left[p - c - kr\right]$$

For price dependent demand $q' < 0$ this implies

$$p(r) = c + kr$$

i.e. marginal cost pricing, which should not come as a surprise. Recall that with demand rectangular the price system p(r) was irrelevant to welfare optimization.

For linear demand functions an elementary proof can be given

$$W = \text{Max} \int_0^R \left( \frac{1}{2}[a - p(r)]^2 + [a - p(r)][p(r) - c - kr] \right) \rho(r) dr - F$$

completing the square under the integral

$$\int_0^R \frac{1}{2} \left( [a - p + p - c - kr]^2 - \frac{1}{2}[p(r) - c - kr]^2 \right) \rho(r) dr$$

$$\text{Max } W = \text{const.} - \frac{1}{2} \underset{p(r)}{\text{Min}} \int [p(r) - c - kr]^2 \rho(r) dr$$

implies $p(r) = c + kr.$

A practical difficulty with marginal cost pricing, when production costs are linear, is that it leaves the firm with a loss equal to fixed costs. As a second best solution we now consider a welfare maximizing price system constrained to let the firm break even.

With a Lagrange multiplier $\lambda$ consider

$$\underset{p(r)}{\text{Max}} \int_0^R \left( \frac{1}{2}[a - p(r)]^2 + [a - p(r)][p(r) - c - kr] \right) \rho(r) dr$$

$$- F + \lambda \left\{ \int_0^R [a - p(r)][p(r) - c - kr] \rho(r) dr - F \right\}$$

solved by

$$p(r) = \frac{\lambda}{1 + 2\lambda} a + \frac{1 + \lambda}{1 + 2\lambda} [c + kr] \tag{20}$$

and the optimal market radius is the maximal one since under this pricing rule quantity demanded then falls to zero at

$$R = \frac{a - c}{k} \tag{21}$$

When fixed cost is absent there is no constraint so that $\lambda = 0$, then $p(r) = c + kr$ and marginal cost pricing results. When the constraint is at its tightest so that $\lambda \to \infty$ then

$$p(r) = \frac{a + c + kr}{2}$$

the perfectly discriminating price system emerges. For mildly constraint problems the pricing rule is a weighted average of these two extremes.

**Mathematical Note 15.7**
**Welfare Comparison of Spatial Price Systems**

To compare consumer welfare under uniform and mill pricing we equalize sales by setting once more

$$p_U = p_M + k\bar{r} \tag{1}$$

for an arbitrary mill price $p_M$. Now since consumers' surplus is the area between the demand curve and the price line, for linear demand functions

$$C_U = \int \frac{1}{2}[a - p_M - k\bar{r}]^2 \rho(r)dr$$

$$C_M = \int \frac{1}{2}[a - p_M - kr]^2 \rho(r)dr$$

Standardizing

$$\frac{1}{2}\int\left[(1-\bar{r})^2 - (1-r)\right]\rho(r)dr$$

$$= \frac{1}{2}\int\left[2(r-\bar{r}) + \bar{r}^2 - r^2\right]\rho(r)dr$$

$$= \frac{1}{2}\int\left(\bar{r}^2 - r^2\right)\rho(r)dr$$

$$= \frac{1}{2}\left[\left(\frac{\int r\rho(r)dr}{\int \rho(r)dr}\right)^2 \int \rho(r)dr - \int r^2\rho(r)dr\right]$$

$$= \frac{1}{2}\frac{1}{\int\rho(r)dr}\left[\left(\int r\rho(r)dr\right)^2 - \int r^2\rho(r)dr \cdot \int \rho(r)dr\right]$$

$$< 0 \qquad \text{by Schwartz's inequality}$$

(cf Beckmann 1976).

For equal sales the consumers' surplus is thus larger under mill pricing than under uniform pricing. This is presumably true for all identical demand functions, but the proof would be more complicated.

*Social surplus.* The social surplus, the sum of consumers' surplus and profits is a more comprehensive welfare measure. Assume profit maximizing prices. Then with identical linear demand functions, profits (and sales) are equal under both pricing systems and the difference of social surplus becomes that of the consumers'surplus.

Therefore mill pricing is also superior when comparing the social surplus under profit maximization.

The consumers' surplus under *discriminatory pricing* is

$$C_D = \int_0^R \frac{1}{2}\left[a - \frac{a+c+kr}{2}\right]^2 \rho(r)dr \tag{2}$$

The consumers' surplus under mill pricing is

$$\int_0^R \frac{1}{2}\left[a - \frac{a+c-k\bar{r}}{2} - kr\right]^2 \rho(r)dr$$

$$C_M = \frac{k^2}{8}\int_0^R \left[\frac{a-c}{k} + \bar{r} - 2r\right]^2 \rho(r)dr \tag{3}$$

Comparing

$$C_M - C_D = \frac{k^2}{8}\int_0^R \left(\left[\frac{a-c}{k} + \bar{r} - 2r\right]^2 - \left[\frac{a-c}{k} - r\right]^2\right)\rho(r)dr$$

and standardizing

$$C_M - C_D = \frac{1}{8}\int_0^R \left[(1+\bar{r}-2r)^2 - (1-r)^2\right]\rho(r)dr$$

$$= \frac{1}{8}\int\left[1 + 2\bar{r} + \bar{r}^2 - 4r - 4r\bar{r} + 4r^2 - 1 + 2r - r^2\right]\rho(r)dr$$

$$= \frac{1}{8}\left[2\bar{r} + \bar{r}^2 - 4\bar{r} - 4\bar{r}^2 + 2\bar{r}\right]M(R) + \frac{3}{8}\int r^2\rho(r)dr$$

$$= \frac{3}{8}\left[\int r^2\rho(r)dr - \frac{\left(\int r\rho(r)dr\right)^2}{\int \rho(r)dr} M\right]$$

$$= -\frac{3}{8}\frac{1}{M}\left[\left(\int r\rho(r)dr\right)^2 - \int r^2\rho(r)dr\int\rho(r)dr\right] > 0$$

for given R, by Schwartz's inequality.

Now assume constant population density and a linear demand function. Observe that the (standardized) market radius for discriminatory pricing is $R_D = 1$ and for mill pricing $R_M = \frac{3}{4}$. Discriminating pricing yields a total social surplus of

$$W_D = \frac{\pi\rho}{2}\int_0^1 (1-r)^2 r dr$$

since consumers surplus and profits are here equal. The total welfare in its market area under mill pricing is

$$W_M = \frac{\pi\rho}{4}\int_0^{\frac{3}{4}}\left[\left(1+\bar{r}-2r\right)^2+\left(1-\bar{r}\right)\left(1-\bar{r}\right)\right]rdr$$

$$= \frac{\pi\rho}{256},$$

compared to

$$W_D = \frac{\pi\rho}{2}\int_0^1 r-2r^2+r^3dr = \frac{\pi\rho}{2}\left[\frac{1}{2}-\frac{2}{3}+\frac{1}{4}\right] = \frac{\pi\rho}{24}$$

$$> W_M$$

since

$$\frac{1}{84} > \frac{9}{256}$$

(cf Holahan 1975).

### Mathematical Note 15.8
### Efficient Size and Structure of the Monocentric City

At what city size (radius R) is the value added by production in a monocentric city maximized? Let the city's industrial production located in the city center (CBD) be produced from labor according to a Cobb Douglas production function $L^\alpha$ and sold in competitive markets at a price p. There is a national labor market in which wages w are set, net of housing and transportation allowances, under competition or collective bargaining. From the net value of output minus net wages must be deducted as further labor costs, the cost of housing and of transportation to work.

Let each household live in one housing unit so that the density of households is the density of the housing units. Costs per area of h housing units per area is assumed to be $bh^\eta$, $\eta > 1$. In a monocentric city of radius R total housing and transportation costs are then

$$\int_0^R\left[bh(r)^\eta + khr\right]2\pi rdr$$

In the optimally sized city the net value generated is maximized

$$\underset{h(r),R,L}{\text{Max}}\ pL^\alpha - wL - \int_0^R\left[bh^\eta + khr\right]2\pi rdr \tag{1}$$

subject to the housing constraint

$$\int_0^R h(r)2\pi rdr \ge L \tag{2}$$

Maximizing the Lagrangean

$$pL^{\alpha} - wL - \int_0^R \left[bh^{\eta} + khr\right] 2\pi r dr + \lambda\left[\int_0^R h 2\pi r dr - L\right] \tag{3}$$

with respect to L

$$0 = \alpha p L^{\alpha-1} - w - \lambda \tag{4}$$

and with respect to R

$$0 = -bh(R)^{\eta} - kh(R) + \lambda h(R) \tag{5}$$

and (under the integral) with respect to h

$$0 = \left[- b\eta\, h^{\eta-1} - kr + \lambda\right] 2\pi r \tag{6}$$

Now (4) is the marginal productivity principle applied to the full labor cost of $w + \lambda$, where $\lambda$ is the housing plus transportation allowance or city bonus to be paid to each worker employed. (6) determining this bonus states that it is equal to transportation cost plus the marginal cost of housing in every location, meaning that the rent should be set at marginal cost. (5) states that a second condition at the city boundary implies, in conjunction with (6) and $\eta > 1$ that

$$h(R) = 0.$$

Hence

$$\lambda = kR \tag{8}$$

and so, using (6),

$$h(r) = \frac{1}{b\eta}\left[\lambda - kr\right]^{\frac{1}{\eta-1}} = \frac{1}{b\eta}\left[k(R - r)\right]^{\frac{1}{\eta-1}} \tag{9}$$

Housing density should fall off as a power function from $h(0) = \dfrac{(kR)^{\gamma}}{b\eta}$ to $h(R) = 0$

abbreviating

$$\gamma = \frac{1}{\eta - 1}$$

The number of households in the city is now

$$\int_0^R h 2\pi r dr = 2\pi \int_0^R \frac{k^{\gamma}}{b\eta}\left[R - r\right]^{\gamma} r dr$$

$$= 2\pi \frac{k^{\gamma}}{b\eta} \int_0^R t^{\gamma}\left[R - t\right] dt\ t$$

$$= \frac{2\pi k^{\gamma}}{b\eta(1+\gamma)(2+\gamma)} R^{\gamma+2} \tag{10}$$

When each household supplies an average of m workers

$$L = \frac{2\pi k^{\gamma} m}{(b\eta)^{\gamma}(1+\gamma)(2+\gamma)} R^{\gamma+2} \tag{11}$$

The marginal productivity principle (4) states now that

$$\alpha p \left[\frac{2\pi m k^{\gamma}}{b\eta(1+\gamma)(2+\gamma)}\right]^{\alpha-1} R^{(\gamma+2)(\alpha-1)} = w + kR \tag{12}$$

using (8), or

$$c_o R^{(\gamma+2)(\alpha-1)} = w + kR \qquad \text{(say)}$$

which may be solved graphically (fig 15.8).

The industry (industries) of the city may even have increasing returns to scale $\alpha > 1$ provided

$$(\alpha-1)(\gamma+2) = (\alpha-1)\left[\frac{1}{\eta-1}+2\right]$$

$$= (\alpha-1)\frac{2\eta-1}{\eta-1} < 1 \tag{13}$$

Since $\frac{2\eta-1}{\eta-1} < 1$ decreases with $\eta$ this condition allows returns to scale to be larger the greater the exponent of the housing cost function $\eta$ is.

Notice that this calculation has been entirely in terms of costs and output value. It pays no attention to the quality of life in the city, the variety of products or services offered and the utility of space and amenities. As wages or the rate of transportation

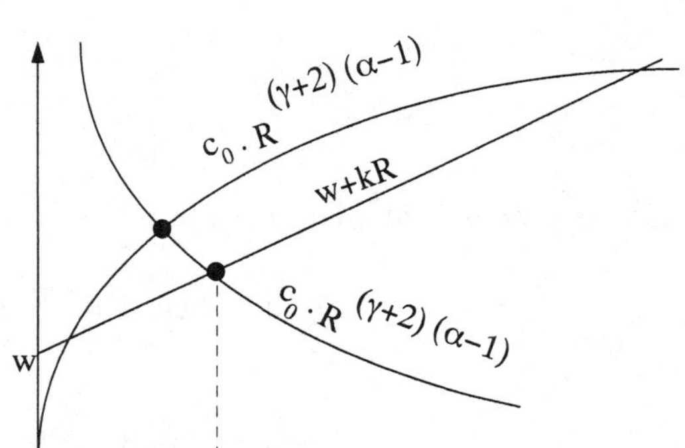


**Fig. 15.8.** Determination of the optimal city radius (for $\alpha > 1$ and $\alpha < 1$)

costs rise ceteris paribus the efficient city size shrinks. One may want the cost of transportation to include the cost of distribution of all city goods and services, such as water, gas, electricity, fire and police protection, ambulance service etc. which are supplied to the residents of the city from the city center. Costs of services that do not depend on city size or that are proportional to city size and hence constant per household, may be ignored.

### Mathematical Note 15.9
### Tolls for Road Investment

Let travel times (cost) t depend on volume x and capacity c as follows

$$t = t\left(\frac{x}{c}\right) \tag{1}$$

Toll income from marginal minus average cost charges is then

$$\frac{x}{c} t'\left(\frac{x}{c}\right) \cdot x \tag{2}$$

Let h be the maintenance and prorated road construction costs per unit of capacity on a particular link (for convenience we have dropped the subscripts $_{ij}$)

The optimal capacity which minimizes costs of traffic plus roads is then determined by

$$\min_c x\, t\left(\frac{x}{c}\right) + hc$$

yielding

$$0 = -x\, t'\left(\frac{x}{c}\right)\frac{x}{c^2} + h \tag{3}$$

or

$$\left(\frac{x}{c}\right)^2 t'\left(\frac{x}{c}\right) = h$$

(say)

$$\frac{x}{c} = g(h) \tag{4}$$

implying that the ratio of transport volume to capacity should depend on road costs h.

Now let all toll income be used for road construction and maintenance. Using (2)

$$h\, c = \frac{x^2}{c}\, t'\left(\frac{x}{c}\right) \tag{5}$$

which is identical to (3): Using tolls for road investments is optimal.

## Conclusion

*The spatial structure of the economy: Ideals and realities.*

"The Spatial Order of the Economy" was the title of Lösch's book of 1940 in the original. At a time when planning was in favor, he emphasized the ordering forces of a market economy which, if not perfectly competitive, could be seen operating at least in "monopolistic competition" with free entry and exit.

From the perspective of today: how orderly is a spatial economy in which market forces are allowed to work out their way? Such fundamental questions tend to be overlooked in the preoccupation with technical questions that prevails in much of location theory.

Put another way; beyond the market failures discussed above, where can one see a case for regulation or even planning to correct undesirable patterns of behavior that arise when private and social cost differ in a spatial context?

Here is an incomplete list of complaints that contributes to our discontent with modern urban life.

- The high cost of urban living, visible in high taxes and land rents
- The long times spent in commuting, aggravated through urban congestion
- The desolate state of cultural amenities
- The limited choice of available services outside high density metropolitan areas
- The need to move to locations, not otherwise attractive, in search of a job
- The dismal picture of the urban landscape and sad state of urban aesthetics
- The fragility of man's environment, not just the natural but also the man-made one
- The instability of neighborhoods, and the destruction or endangering of urban landmarks by economic growth and change.

For example, creating beauty in the city through attractive buildings, decoration and display, care of gardens, etc, is not rewarded while creating ugliness through attention grabbing displays and advertising is not in any way penalized.

The case of traffic congestion is also notorious: putting one's own vehicle into a congested stream costs only the average prevailing delay but not the marginal additional delay to all other vehicles (see 13.2).

Some problems are clearly untractable except by huge subsidies, such as the dearth of services in areas of low population density, a result simply of economies of scale (or high fixed cost) in the production of most goods and services.

It has been proposed to do away with all cities or spatial concentration and have population and all economic activities spread out at low densities, but this would force us to forego either the advantages of the division of labor on which developed economies rely or incur unconscionably high burdens of transportation costs.

What is indicated instead are corrections of the undesirable consequences of market forces of monopoly and monopolistic competition in spatial contexts through cost correction, charges, taxes or subsidies, and when all else fails, attempts at regulation through urban and regional planning.

These are important topics in their own right that exceed the domain of location theory proper, which is after all: the study of locational choices by household and enterprises in the context of spatial markets and of the resulting structures in a spatial economy.

## References

Albegov M, Anderson A, Snickers F (eds) (1982) Regional Development Modelling: Theory and Practice. North-Holland, Amsterdam
Alden J, Morgan R (1974) Regional Planning: A Comprehensive View. Wiley, New York
Beckmann MJ (1976) Spatial Price Policies Revisited. Bell Journal of Economics 7, 2, Autum:619–630
Blunden WR (1984) The Land Use/Transportation Systems. Pergammon Press, Sydney
Boudeville JR (1968) L'Espace et les pole de croissance. Presses Universitaire, Paris
Doxiades KA (1968) Ecistics. Oxford University Press, New York
Glassmeier A (1983) Regional Planning and Economic Development: A Bibliography. CPC Bibliographies, Chicago
Gillie FB (1967) Basic Thinking in Regional Planning. The Hague, Mouton
Hichs JR (1956) Value and Capital. Cambridge: Cambridge Univ. Press
Holahan WL (1975) The Welfare Effect of Spatial Price Discrimination. AER 65, 3:498–513
Holcombe RG (1995) Public Policy and the Quality of Life: Market Incentives Versus Government Planning. Greenwood Press, Westport, Ct.
Kuklinski A (1974) Regional Information and Planning. Mouton, Paris
Leven C (1970) An Analytical Framework for Regional Development Policy. MIT Press, Cambridge
Marshall JM (1995) Services and Space. Wiley, New York
Masser J (1972) Analytic Models for Urban and Regional Planning. Wiley, New York
Paelinck J (ed) (1984) The Quest for Optimality. Gower Publications, Aldershot, Hamsphire

# Suggested Reading


Dean RD, Leahy WH, McKee DC (eds) (1970) Spatial Economic Theory. Free Press, New York

Greenhut M (1995) Location Economics: Theoretical Underpinnings and Applications. Edward Elgar, Aldershot

Hoover EM (1971) An Introduction to Regional Science. Alfred A. Knopf, New York

Isard W (1975) Introduction to Regional Science. Prentice-Hall, Englewood Cliffs

Krugman P (1993) Geography and Trade. MIT Press, Cambridge

Lösch A (1954) The Economics of Location. Yale University Press, New Haven

McCann R (1998) The Economics of Industrial Location. Springer-Verlag, Heidelberg

Ponsard C (1983) History of Spatial Economic Theory. Springer-Verlag, Heidelberg

Puu T (1997) Mathematical Location and Land Use Theory. Springer-Verlag, Heidelberg

Thisse JF, Button KJ, Nijkamp P (eds) (1996) Location Theory, Colume I. Edward Elgar, Cheltenham

van der Bergh JCJM, Nijkamp P, Rietveld P (eds) (1996) Recent Advances in Spatial Equilibrium Modelling. Sringer-Verlag, Heidelberg

Druck: Strauss Offsetdruck, Mörlenbach
Verarbeitung: Schäffer, Grünstadt